APPLIED BIOCATALYSIS

VOLUME 1

APPLIED BIOCATALYSIS

VOLUME 1

EDITED BY

HARVEY W. BLANCH

DOUGLAS S. CLARK

*University of California at Berkeley
Berkeley, California*

Marcel Dekker, Inc.   **New York · Basel · Hong Kong**

Library of Congress Cataloging--in--Publication Data

Applied biocatalysis / [edited by] Harvey W. Blanch, Douglas S. Clark
   p. cm.
  Includes bibliographical references and index.
  ISBN 0-8247-8533-9
  1. Enzymes-- --Industrial applications. I. Blanch, Harvey W.
  II. Clark, Douglas S.
  TP248.E5A66 1991
  660'.634--dc20

                                        91–17508
                                          CIP

This book is printed on acid-free paper.

MARCEL DEKKER, INC.
270 Madison Avenue, New York, New York 10016

Current printing (last digit):
10 9 8 7 6 5 4 3 2 1

PRINTED IN THE UNITED STATES OF AMERICA

Preface

Applied Biocatalysis, Volume 1, begins a new series that will highlight recent advances in the application and commercialization of biological catalysis. The importance and timeliness of this series are self-evident in view of revolutionary developments in protein chemistry and the expanding interest of industry in the use of enzyme technology for the production of food products, pharmaceuticals, specialty chemicals, and fuels. The redesign of enzymes by mutagenesis techniques and the use of enzymes under novel conditions (e.g., in nonaqueous media) are but two examples of breakthrough developments that point to an era of unprecedented growth in enzyme technology. Within the broad scope of biotechnology, enzyme technology also promises to play a growing role in the translation of advances in molecular biology into successful products and processes. *Applied Biocatalysis* will chronicle this exciting evolution and will present comprehensive reviews of important achievements in biocatalytic science and technology.

In this first volume, we are fortunate to have assembled contributions from world-class authorities on enzyme function in nonaqueous solvents and applications in synthesis. Key topics addressed in this volume include nonaqueous enzymology in general and protein design for organic solvents in particular, and enzymatic synthesis of carbon-carbon bonds, novel oligosaccharides, and modified steroids. The chapters are practical in approach and should be valuable to experienced as well as potential users of enzymes. Educators should also find this book useful as a source of both general information on enzyme technology and specific examples of enzyme catalysis.

Much appreciation is due to the authors for their generous cooperation. We would also like to acknowledge the members of the Editorial Board (Charles Craik, J. Bryan Jones, Alexander Klibanov, Peter Schultz, and Johannes Tramper) for their invaluable assistance throughout this project.

Harvey W. Blanch
Douglas S. Clark

Contents

Contributors

Frances H. Arnold   *Department of Chemical Engineering, California Institute of Technology, Pasadena, California*

Mark David Bednarski   *Department of Chemistry, University of California at Berkeley, Berkeley, California*

Jonathan S. Dordick   *Department of Chemical and Biochemical Engineering, University of Iowa, Iowa City, Iowa*

Jack Y. Hwang   *Department of Chemical Engineering, California Institute of Technology, Pasadena, California*

Kurt G. I. Nilsson   *Chemical Center, University of Lund, Lund, Sweden*

Sergio Riva   *Istituto di Chimica degli Ormoni, Consiglio Nazionale delle Ricerche, Milan, Italy*

APPLIED BIOCATALYSIS

VOLUME 1

1

Principles and Applications of Nonaqueous Enzymology

Jonathan S. Dordick
University of Iowa, Iowa City, Iowa

INTRODUCTION

Nature is extremely diverse in terms of the large number and many types of organic molecules required for life. This diversity is solely due to the wide catalytic scope of enzymes. It is the ability to harness the catalytic power of enzymes and use it for the synthesis of commercially important products that represents the core technology of applied biocatalysis. Enzymes are highly selective catalysts that operate typically under mild reaction conditions (e.g., ambient temperatures and pressures, neutral solutions). Such properties have enabled enzymes to become valuable catalysts in the food, beverage, and diagnostics industries. The same properties make enzymes potentially attractive in synthetic chemistry, especially in the synthesis of pharmaceuticals, chiral intermediates, specialty polymers, and biochemicals.

The chemical industry, however, has been slow to employ enzymes. Perhaps the most significant reason for this has been the strict adherence by both chemists and biochemists to the conventional notion that enzymes function only in aqueous solutions. Indeed, it is stated in virtually every biochemistry textbook that enzymes are aqueous-based and require water for activity and that organic solvents, with few exceptions, only serve to destroy enzyme function. This is unfortunate because water, while an ideal solvent for the predomi-

Table 1 Potential Advantages of Employing Enzymes in Organic Media

---

1. Increased solubility of nonpolar substrates.
2. Shifting thermodynamic equilibria to favor synthesis over hydrolysis.
3. Suppression of water-dependent side reactions (e.g., hydrolysis of acid anhydrides and halides, polymerization of quinones.
4. Alteration in substrate specificity.
5. Immobilization is often unnecessary because enzymes are insoluble in organic solvents.
6. Enzymes may be recovered by simple filtration or centrifugation.
7. If immobilization is required for optimal flow considerations, simple adsorption onto nonporous surfaces (e.g., glass beads) is satisfactory. Enzymes are unable to desorb from these surfaces in nonaqueous media.
8. Ease of product recovery from low boiling, high vapor pressure solvents.
9. Enzymes exhibit enhanced thermostability.
10. Elimination of microbial contamination.
11. Potential enzymes to be used directly in a chemical process.

Source: Ref. 3.

nantly polar species required for life (e.g., amino and nucleic acids, carbohydrates, proteins, cofactors), is a poor solvent for nearly all applications of industrial chemistry. Most organic compounds of commercial interest are very sparingly soluble and often unstable in aqueous solutions. Hence, industrial chemistry is based on organic solvents and has mostly ignored the potential benefits of highly selective enzymic catalysts. This situation is quickly changing. The application of enzymes in nonaqueous media enables biocatalysts to compete successfully with traditional chemical catalysts in a variety of synthetic processes. The notion that enzymes are active only in water is fast becoming obsolete.

Indeed, it is ironic that many enzymic processes in the cell do not take place in a true aqueous environment, and that these processes would not be optimal in such environments. Many enzymes or multienzyme complexes, including lipases, esterases, dehydrogenases, and cytochromes, function in natural hydrophobic environments, usually in the presence of or immobilized to a membrane [1,2]. The water activity under these conditions is significantly less than unity, and the concentration of water is well below the 55 M value found in aqueous solutions.

From a biotechnological standpoint, there are numerous potential advantages in employing enzymes in organic media (Table 1) [3,4]. These advantages are attractive in selective chemical synthesis, and a number of approaches using enzymes in nonaqueous environments have been conceived. For example, water-miscible organic solvents in relatively low concentrations (e.g., < 30% v/v) have been used to increase the solubility of hydrophobic substrates [5]. Soluble enzymes have also been used in the aqueous components of both biphasic aqueous-organic systems [6-8], and inside reverse micelles dissolved in nonpolar solvents [9,10]. The enzyme in each of these three solvent designs, however, resides in a predominantly aqueous environment.

Several of the properties listed in Table 1 are impossible unless the bulk water phase is eliminated. Such a situation exists using enzymes in monophasic organic solvents [3]. These systems are defined as those that lack a distinct aqueous phase with an insoluble enzyme catalyst (enzymes are insoluble in nearly all organic solvents) suspended in nearly anhydrous (no added water) media or water-miscible cosolvents employing the organic solvent as the predominant system component. Unlike the other solvent reaction systems, enzymes in monophasic solvents are not in direct contact with a bulk aqueous phase and, therefore, all the potential advantages listed in Table 1 have been realized. This chapter first focuses on the principles of catalytic function of enzymes in monophasic organic media, the consequences of removing bulk water from an enzymic reaction mixture, the mechanistic and kinetic behavior of enzymes in nonaqueous media, and the factors that must be taken into account for choosing the proper solvent and enzyme preparation for a given reaction. A description of applications of commercially relevant enzymes in monophasic organic solvents follows this discussion.

ROLE OF WATER IN ENZYMATIC CATALYSIS

The chemical and physical properties an enzyme exhibits largely depend on the direct or indirect role of water in all noncovalent interactions [11,12]. These interactions include electrostatic, van der Waals, and hydrophobic forces, and hydrogen bonding, all of which help to maintain the catalytically correct enzyme conformation. Replacing water with an organic solvent would, in principle, distort the native structure of an enzyme molecule and surely lead to inactivation. Indeed, no truly anhydrous enzyme preparation has catalytic activity, and some water is necessary for catalysis. Because of the importance of water in enzyme function, it is necessary to review the role of water in enzyme structure and function, and to describe the consequences of removing bulk water from an enzyme reaction mixture.

Essential Water

While the need for water is well understood, the amount required has only recently been investigated. The most comprehensive set of studies was performed using lysozyme and was based on the ability of the dehydrated protein to rehydrate, via equilibration with humid air [13-15]. Four distinct hydration levels were shown to exist. Up to 0.07 g H_2O per gram of lysozyme, the enzyme can be considered to be devoid of "wet" or bulk water, and that water predominantly interacts with the protein's charged functional groups. The water mobility is only 1% as high as that in bulk solution. Heat capacity measurements suggest that the water structure at this level of hydration is between the levels of ice and water. Between 0.07 and 0.25 g H_2O per gram of lysozyme, a shift in the water distribution on the surface of the protein occurs from purely ionic sites to polar regions, and the enzyme becomes more mobile. Between 0.25 and 0.38 g H_2O per gram of lysozyme, enzyme activity is observed and the water associated with the lysozyme approaches monolayer coverage; the polar sites of the protein are hydrated first, followed by the nonpolar regions. Above 0.38 g H_2O per gram of lysozyme ($\approx$ 300 molecules H_2O per enzyme molecule), the protein is fully hydrated and any additional water leads to a situation similar to what is observed when the enzyme is placed in bulk water. Above 50% monolayer coverage, protein structure is nearly identical to that found in bulk solution. Hence, very little water is actually required for native enzyme structure and function to be realized.

This work has been extended to enzymes in organic solvents. Chymotrypsin suspended in octane requires roughly 50 molecules of water per molecule of enzyme for activity to be observed [16]. Other hydrolytic enzymes also appear to require low amounts of water for catalytic activity in organic media [17-19]. It appears, however, that lysozyme, chymotrypsin, and several other hydrolases may be the exception rather than the rule in that very few molecules of water are required for activity. In most cases, water must be supplemented to the reaction medium for catalysis to occur. For example, polyphenol oxidase requires 0.5% (v/v) aqueous buffer in chloroform [20], and horseradish peroxidase requires 0.25% (v/v) aqueous buffer in toluene for optimal catalytic function [21]. In both cases the added water entrained in the insoluble enzyme particles and no separate aqueous phase results.

Significant differences in enzyme activity are also observed in various solvents of different hydrophobicities. Recent studies using three unrelated enzymes—polyphenol oxidase, alcohol dehydrogenase, and alcohol oxidase—revealed that significant differences exist between observed catalytic activity of an enzyme and the concentration of water in several organic solvents [22]. For example, in ethyl ether containing 0.6% water, alcohol oxidase was more than 3 orders of magnitude more active than in the more polar *sec*-butanol containing the same concentration of water. Polyphenol oxidase in hexyl acetate containing 0.5% added water was nearly 10% as active as in fully aqueous solution, whereas in *n*-butanol supplemented with 0.5% water, no activity was observed. Clearly, enzymes suspended in hydrophobic media require less water to support catalytic activity than those suspended in hydrophilic solvents.

The cause for this behavior may be the propensity of the essential water adsorbed to a protein molecule to favorably partition into hydrophilic solvents, resulting in a loss of catalytic activity (see below: "Effect of Organic Solvents on Biocatalysis"). Most reactions described in the literature that employ enzymes in organic media are reported with a given level of water in the reaction mixture. Instead of correlating enzyme activity to *total* water in the reaction, a more appropriate correlation might be based on actual water adsorbed to the enzyme. Only when the aforementioned enzyme activities are plotted as a function of the water bound to the enzyme (Fig. 1) do individual differences between enzymes and solvents disappear and level off at approximately 0.3 g H_2O per gram of enzyme. Hence, it is the water adsorbed to the enzyme that is critical in maintaining enzyme activity, not the total water in the solvent. The structural features that determine the propensity of water to stay bound to an enzyme remain to be elucidated.

Adsorbed Versus Bulk Water

The changes in the physicochemical properties of lysozyme powders and the sensitivity of enzymes to the level of hydration of the enzyme itself suggest that water associated with a protein molecule is significantly different from that in bulk solution. Studies of proton relaxation time of water bound to an inert surface indicate that water adsorbed directly onto the surface of a protein has restricted motion relative to water in the bulk and that subsequent monolayers are much less restricted [23]. Such restricted motion in the water monolayer leads to reduced density, greater viscosity, reduced ability to participate in hydrogen bonding, and a higher degree of hydrophobicity as compared with bulk water [24–26]. This may result in thermodynamic relationships that are different from bulk aqueous solution. For example, recent studies performed with adsorbed water on cellulose acetate films indicate that MgATP formation from MgADP and inorganic phosphate is possible, whereas in bulk water, ATP hydrolysis is favored [27].

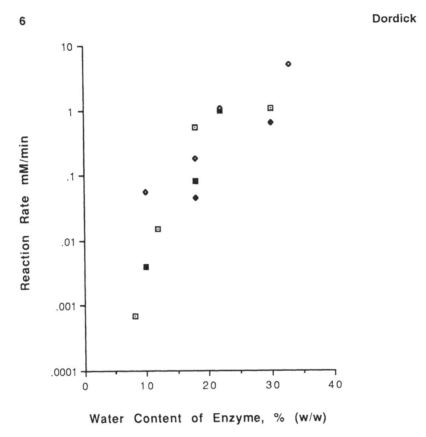

Figure 1 Correlation between yeast alcohol oxidase activity and the amount of water bound to the enzyme: ⊡ , ethyl acetate; ◇ , *tert*-amyl alcohol; ■ , diethyl ether; ◇ , *sec*-butanol; ◆ , butyl acetate. (From Ref. 22)

Replacement of Water by Solvent Mimics

The differences between adsorbed and bulk water, particularly the increase in hydrophobicity and the decrease in hydrogen bonding character, indicate that water may be replaced by solvents with physicochemical properties intermediate between water and hydrophobic organic solvents without significantly affecting catalytic activity. This would be particularly so if such solvents had hydrogen bonding characteristics, as was shown to be true in several specific studies employing water mimics such as formamide, ethylene glycol, and glycerol. The presence of 1% formamide in butyl acetate containing 0.4% water increased the activity of alcohol dehydrogenase 15-fold [22]. More dramatic activity enhancements were observed in the thermolysin-catalyzed synthesis of peptides in *tert*-amyl alcohol [28]. In the presence of 1% water and 9% ethylene glycol, 200-fold rate enhancements were observed over the reaction containing only 1% water. Reducing water concentration in a reaction mixture is often advantageous. In the case of thermolysin-catalyzed peptide synthesis, unwanted hydrolysis of the newly formed peptides is eliminated in the presence of a water mimic such as ethylene glycol [28].

The preceding studies indicate that only a very small fraction of bulk water is required for enzyme structure and function. It should be possible to keep an enzyme fully hydrated and active, with water comprising only a small proportion of the reaction volume and the bulk reaction volume consisting of an organic solvent. This is the overriding principle of enzymatic catalysis in monophasic organic environments, and recent work in several laboratories around the world has supported it [3,4,29].

EFFECT OF ORGANIC SOLVENTS ON BIOCATALYSIS

Organic solvents affect the nature of biocatalysis in several ways, including interaction with the essential hydration level of the enzyme, direct participation in the reaction mechanism, alteration of protein structure and flexibility, and alteration of observed enzyme kinetics.

Macroscopic Effects

The presence of an organic solvent as a reaction medium has several ramifications.

Organic Solvent Interaction with Essential Water

The organic solvent may affect enzyme structure and function by directly interacting with the enzyme's essential level of hydration. Specifically, highly polar organic solvents re capable of solubilizing large amounts of water and will tend to "strip" away from the enzyme the water necessary for catalytic structure and function [21]. Conversely, hydrophobic solvents are less able to solubilize the adsorbed water and are less likely to distort the enzyme structure and activity. This scenario is well illustrated with horseradish peroxidase, an enzyme particularly susceptible to solvent hydrophobicity (Table 2). Direct measurement of bound water to an enzyme in organic media with Fourier transform infrared spectroscopy has confirmed water stripping from an enzyme by hydrophilic solvents through a shift in the frequency of protein amides [30].

Organic Solvent Interaction with Diffusible Substrates or Products

Another way solvents affect enzymatic activity is by interacting with diffusible substrates or products of the reaction. For example, chloroform (a good phenoxy radical quencher) significantly reduced the activity of horseradish peroxidase–catalyzed polymerization of phenols [21,31], a reaction initiated by the one-electron oxidation of the phenol to a phenoxy radical.

Solvent-Mediated Inactivation via Direct Interaction with Enzyme

The solvent can also inhibit or inactivate the enzyme via direct interaction with the protein. In this case, the solvent alters the native conformation of the pro-

Table 2   Horseradish Peroxidase Catalyzed Oxidation of *p*-Anisidine in Monophasic Organic Solvents

Solvent	Reaction rate (μmol/mg enzyme/min)[a]
Toluene	60
Benzene	54
Hexane	75
Hexadecane	75
Cyclohexane	75
Methylene chloride	9
Chloroform	9
Ethyl acetate	45[b]
Butyl acetate	62[b]
Diethyl ether	105[b]
2-Butanone	23[b]
n-Amyl alcohol	11[b]
Diemthyl sulfoxide	0[c]
Methanol	0[c]
Dioxane	0[c]
Dioxane + 5% aqueous buffer	8
Dioxane + 15% aqueous buffer	33
Dioxane + 30% aqueous buffer	129

[a]Reaction conditions: 1 mM *p*-anisidine, 0.2 mM H_2O_2, 1 μg/mL peroxidase, 0.25% v/v phosphate buffer, pH 7.
[b]Solvents were saturated with phosphate buffer prior to use.
[c]No reaction observed with 1% v/v phosphate buffer.
Source: Ref. 21.

tein by disrupting hydrogen bonding, and ionic and hydrophobic interactions. This type of inactivation is particularly severe when the enzyme is soluble in the organic solvent. Dimethylformamide, dimethyl sulfoxide, and 2-chloroethanol are well-known solvents capable of solubilizing enzymes [32]. Hydrogen bonding, as well as ionic and hydrophobic interactions both within the enzyme molecules and between the enzyme and the solvent, will be considerably different from the aqueous solution case. The unfolded form of the enzyme becomes thermodynamically favored over the native conformation [33,34]. For example, ribonuclease dissolved in increasing concentrations of 2-chloroethanol undergoes a transition from the native state to a partially unfolded protein, to a newly folded, incorrect structure [35]. In extreme cases, the hydrophobic "core" of enzymes becomes exposed to the organic solvent [36].

Solvents that do not solubilize enzymes may also cause inactivation. Monohydroxylic alcohols are notorious for causing enzyme denaturation. Cytochrome c, myoglobin, and chymotrypsin are severely inactivated in less than 10% v/v *sec*-butanol. Polyhydroxylic alcohols including ethylene glycol and propylene glycol, however, appear to stabilize proteins at low concentrations. In such solvents, enzymes are not inactivated unless greater than 60% v/v solvent is added [37]. In other water-miscible solvents such as dioxane and acetone, enzymes are least stable at or around the concentration of solvent that causes precipitation [38].

At high concentrations of organic solvents, and with few exceptions (see above), enzymes are insoluble. The insolubility of enzymes in high concentrations of organic solvents appears to preclude the enzyme from undergoing unfolding in the presence of the solvent. This is primarily a kinetic effect; the activation energy for protein unfolding in hydrophobic solvents must be high, thereby enabling an enzyme to retain its native structure, albeit in the face of thermodynamic driving forces favoring denaturation [19]. Structural rigidity can be observed by electron spin resonance (ESR) spectroscopy; the more hydrophobic the organic solvent, the fewer rotational degrees of freedom are observed for a spin label attached to an enzyme's active site [39]. Figure 2 depicts the effect of solvent hydrophobicity on the rotational motion of a nitroxide spin label covalently attached to alcohol dehydrogenase. The mean rotational correlation time (T_R) of the spin label increases as solvent hydrophobicity increases, which indicates that the enzyme's active site becomes rigid in apolar solvents [40].

Structural Rigidity of Enzymes in Organic Media

The rigidity of enzymes in organic solvents has led to several interesting phenomena regarding the physicochemical state of the enzyme. One intriguing aspect of enzymatic catalysis in low water environments if the effect of reaction

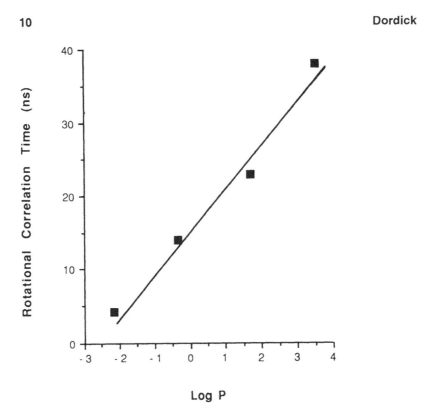

Figure 2   Relationship between mean rotational correlation time (ns) to solvent hydrophobicity using horse liver alcohol dehydrogenase spin labeled with 4-(2-iodoacetamido)-Tempo. (From Ref. 39.)

pH. In reaction systems that lack a distinct aqueous phase, pH is not a measurable quantity. Certainly, there will be a definite concentration of protons associated with the enzyme-bound water; however, this "enzyme pH" cannot be adequately measured. Furthermore, the pH should not correspond to the same physical parameters as in aqueous solutions, given the aforementioned differences between enzyme-bound and bulk water. Regardless of whether the pH of the hydrated enzyme can be effectively determined, enzyme activity depends on the correct ionization state of the enzyme's charged functional groups [19,38,41]. In aqueous buffer, these groups acquire an ionization state that depends on the pH of the solution. In organic solvents, there is no driving force to effect change in the ionization state of charged groups on the protein, and the chemical state of an enzyme remains unchanged from that of the aqueous solution from which the enzyme was last recovered. This has been shown to be the case for a number of enzymes, including lipases and proteases and several oxidoreductases, where pH optima in organic media, obtained by lyophilization of the enzyme from a given pH, are identical to optima in aqueous solutions [19–21,38]. Enzyme stability is also highly pH dependent in organic media. For example, an alkaline protease has been shown to be 20-fold more stable in pyridine at the pH optimum of 9.5 as compared with pH 7 [41a]. From a practical standpoint, pH optimization is vital for effective application of enzymes in solvents. Fortunately, it is also one of the most easily controlled variables.

Another phenomenon associated with the rigidification of enzymes in organic media is alteration in substrate specificity. When the water content of an organic solvent is decreased, enzymes become less flexible and the active sites less amenable to accept bulky substrates. For example, porcine pancreatic lipase in tributyrin will catalyze the transesterification reaction with bulky alcohols, including *tert*-butanol and 3-methyl-3-hexanol, in the presence of 0.7% (v/v) water [42]. In the presence of 0.015% (v/v) water, the lipase is unable to accept these alcohols as acyl acceptors, whereas less bulky primary alcohols remain capable of acting as substrates.

The structural rigidity of enzymes in organic solvents can be used, advantageously, to activate the enzyme for catalysis. The activity of subtilisin in several organic solvents can be improved more than 100-fold by lypophilizing the enzyme in the presence of a competitive inhibitor [43]. The inhibitor acts as a ligand, which activates the enzyme by causing a conformational change in the subtilisin molecule. The structurally altered subtilisin active site is effectively "molded" into the proper configuration for binding to a substrate that is structurally similar to the activating ligand, as would be expected for a ligand acting as a competitive inhibitor of the substrate. The rigidity of enzymes in organic solvents is a prerequisite for this phenomenon. Subtilisin prepared with a ligand is not activated in aqueous solutions, presumably because the native conformation is reformed. A similar ligand "imprinting" phenomenon can be used to convert inert proteins into selective binding agents [44]. Bovine serum albumin (BSA), freeze-dried in the presence of L-tartaric acid, bound the ligand in anhydrous ethyl acetate 30-fold tighter than BSA not imprinted with the ligand or in aqueous solution following imprinting.

Strong ligand binding to proteins in organic media has been shown to occur in antibodies as well [45]. The binding of 4-aminobiphenyl to a monoclonal antibody in acetonitrile was diminished 10-fold from that in water. However, the *selectivity* of hapten binding in acetonitrile was significantly enhanced over that in aqueous solutions. For example, antibody binding of 4-aminobiphenyl in acetonitrile was unaffected by 2-aminobiphenyl, whereas some inhibition of binding was observed in water. The hydrophobicity of the solvent exerted a controlling influence on hapten binding; the more hydro-philic the solvent, the stronger the binding of antigen to antibody.

The rigidity of enzymes in organic solvents provides an opportunity to use solvents as a means of altering substrate specificities and regulating enzyme function. Such a situation is far simpler and less expensive than altering enzyme structure and function using protein engineering. The advantage of protein engineering, however, is that specific changes to a protein can be made independent of reaction conditions.

Thermostability of Enzymes in Organic Media

Enzymes inactivate at high temperatures in aqueous media due to both the partial unfolding of the protein and covalent alterations in the primary structure of the molecule [46]. Water is required in these inactivation processes [47]. In the absence of water, it is expected that enzymes should become more thermostable. This has been recently confirmed in several studies using different enzymes. Lipases have been shown to retain considerable activity at 100°C in tributyrin containing 0.015% (v/v) water; a half-life of 12 hours was obtained, while in aqueous buffer a half-life of seconds is observed [42]. Mitochondrial F_1-ATPase and cytochrome oxidase are active and stable in toluene containing 1.3% water (v/v) at 90°C, and the half-life could be extended nearly 100-fold by lowering the water content to 0.3% [38]. Enzymes in whole cells suspended in "dry" organic solvents also show increased thermostabilities. For example, lipase from *Rhodococcus equi* is highly active at 70°C in alcohols as long as the solvents are dry; but if they are wet, stabilities are poor [49].

Organic Solvent Effects on Enzyme Kinetics

The reaction medium exerts a profound effect on the kinetics of enzymatic catalysis; substrate specificity and catalytic efficiency depend on the ability of the enzyme to utilize the free energy of binding with the substrate [50,51]. This binding energy reflects the difference between binding energies of substrate–enzyme and substrate–solvent interactions [52]. Kinetic parameters describing enzyme function, such as binding constant with substrate K_s, Michaelis constant K_m, and catalytic turnover k_{cat} or V_{max}, therefore, depend strongly on the solvent parameters of the reaction. It may be expected that the replacement of water with an organic solvent would lead to profound changes in the observed kinetics of enzymatic catalysis.

A Brief History

The ability of organic solvents to alter the kinetics of enzymes was well known in the middle part of this century based on several key studies. Kaufman and Neurath investigated the chymotrypsin-catalyzed hydrolysis of N-acetyl-L-tyrosinamide in the presence of low volumes of methanol [53]. Catalytic turnover k_{cat} was unaffected up to 21% (v/v) methanol, while apparent substrate binding was observed to decrease. No attempt was made to correlate physicochemical properties of the organic solvent (e.g., hydrophobicity parameters, dielectric constant) with the observed kinetics. Barnard and Laidler also investigated the effect of methanol on chymotrypsin catalysis, using the hydrolysis of an ester of hydrocinnamic acid as a model substrate [54]. Linear correlations between reaction rate and the inverse of the dielectric constant of methanol–water mixtures were obtained. Both studies indicated that the methanol caused an inhibition of catalytic function. It was suggested that the methanol competed with water for binding sites near or in the chymotrypsin's active site [55]. No direct evidence for either competitive or noncompetitive inhibition was obtained, however.

The effect of high concentrations of dioxane on chymotrypsin catalysis was investigated by Bender, and several interesting kinetic patterns emerged [56]. Above 20% v/v dioxane, the apparent K_m of chymotrypsin-catalyzed hydrolysis of N-benzoyl-L-tyrosine p-nitroanilide was constant and approximately 1000-fold greater than in aqueous buffer. Furthermore, the rate constant of acylation, k_2 (the rate-controlling step in the chymotrypsin-catalyzed hydrolysis of amide substrates and a water-independent reaction [57]), was found to be independent of dioxane concentration. Hence, in a water-independent mechanistic step, the rate-determining process is insensitive to water concentration. Hydrolysis rates of N-acetyl-L-tryptophan p-nitrophenyl ester (k_3, or deacylation, is rate determining for ester substrates [57]), however, were strongly dependent on water concentration; upward of a 30-fold in the deacylation rate constant k_3 was observed in 80% dioxane as compared to aqueous solution. Thus, an aqueous-dependent deacylation step was affected by the concentration of water in the organic solvent reaction system.

While water affected only the deacylation step, catalytic efficiency in the hydrolysis of the amide substrate k_{cat}/K_m (= k_2/K_s for chymotrypsin catalysis) was decreased by more than 3 orders of magnitude solely because of diminished substrate binding. Hence, even when water was not a substrate, catalytic efficiency of the enzyme was affected. Similar results have been obtained in nearly anhydrous organic solvents [16,19]. The transesterification activities of chymotrypsin and subtilisin in octane are at least 3 orders of magnitude lower than corresponding hydrolytic activities in aqueous solution.

Hydrophobicity Effects on Enzyme Function
in Nonaqueous Media

One hypothesis regarding the decrease in catalytic activity in organic media as opposed to water is the partitioning of the substrate from the enzyme's active site to the hydrophobic organic solvent, thereby lowering the binding affinity of substrate with enzyme while increasing the affinity of substrate with solvent. Chymotrypsin and subtilisin have hydrophobic active sites [52]. In water, hydrophobic substrates favorably partition from the bulk aqueous medium into the enzyme active sites. Hydrophobic solvents would diminish this partitioning, leading to an increase in the observed K_s of the substrate and a decrease in the catalytic efficiency of the enzyme toward the hydrophobic substrate. This hypothesis is supported by studies of chymotrypsin-catalyzed hydrolysis of *N*-Ac-L-Trp-ethyl ester in water-miscible solvents [58]. Plots of log (k_{cat}/K_m) versus ΔG_{tr} (defined as the free energy of substrate transfer from water to solvent) are linear, which indicates that a relationship exists between substrate binding and solvent hydrophobicity.

More recently, Dordick and coworkers examined the horseradish peroxidase catalyzed oxidation of phenols in a variety of organic solvents wherein both substrate and solvent hydrophobicities were varied [59]. Peroxidase is an ideal enzyme for such a study because it catalyzes identical reactions in aqueous as well as organic media [60]. In addition, phenols are highly soluble in organic solvents and the individual kinetic constants V_{max} and K_m can be measured.

Phenols with para substituents that differed only in hydrophobicities, not in electron donation or withdrawal, were used and ranged from methoxy- to *tert*-butyl. A variety of water-miscible and -immiscible organic solvents were chosen with different levels of hydrophobicity. Figure 3 depicts the dependence of catalytic efficiency of peroxidase on substrate hydrophobicity π in aqueous buffer and in several organic solvents. In all cases, linear relationships between catalytic efficiency and substrate hydrophobicity were obtained. Furthermore, peroxidase was found to be up to 4 orders of magnitude less efficient in organic media compared to aqueous solution. This substrate effect became more pronounced as the solvent hydrophobicity increased. The slopes of peroxidase activity versus π decreased as solvent hydrophobicity, represented as log P (defined as the logarithm of the partition coefficient of solvent between 1-octanol and water [61]), increased (Fig. 4).

These findings are consistent with the partitioning behavior of hydrophobic phenols from the bulk reaction medium into the peroxidase's active site. This partitioning is likely to diminish as substrate and solvent hydrophobicities increase, thereby necessitating a larger concentration of phenols to saturate the

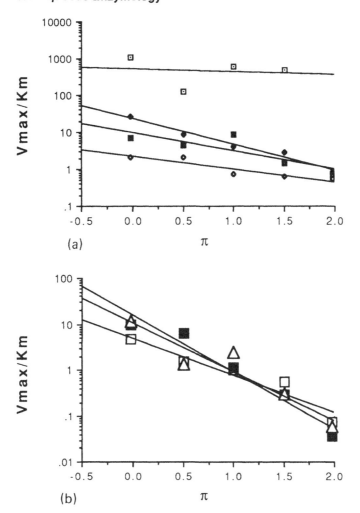

Figure 3 Catalytic efficiency of horseradish peroxidase in aqueous and nonaqueous media. (a) Water-miscible solvents: □ , aqueous buffer (0.01 M phosphate, pH 7); ◆ , dioxane with 30% aqueous buffer; ■ , dioxane with 20% aqueous buffer; ◇ , dioxane with 5% aqueous buffer. (b) Water immiscible solvents: □ , ethyl acetate, containing 2% aqueous buffer;   △ , propyl acetate, containing 1.5% aqueous buffer;   ■ , butyl acetate, containing 1% aqueous buffer.

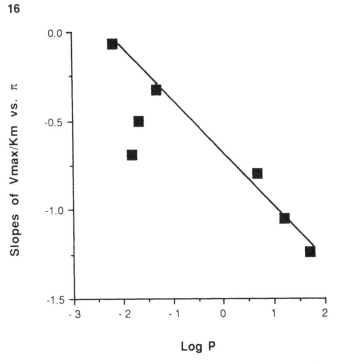

Figure 4   Effect of solvent hydrophobicity on the catalytic efficiency of horseradish peroxidase. Slopes are calculated from Figure 3.

enzyme. This results in a higher apparent substrate K_m in organic versus aqueous media. Similar increases in apparent K_m have been observed for trypsin catalysis in dioxane–water mixtures [62].

Catalytic turnover was *not* necessarily decreased in nonaqueous media [59]. Inspection of Figure 5 reveals that peroxidase turnover in the oxidation of *p*-cresol is *stimulated* in several organic solvents. For example, in 80% dioxane and butyl acetate, *p*-cresol oxidation is more than threefold faster than in aqueous buffer. This further supports the view that solvent hydrophobicity primarily affects the binding of substrates to enzymes. Once this binding has been achieved, however, catalytic activity may remain high.

The high linearity of the relationship between catalytic efficiency and substrate or solvent parameters (π and log P are partition coefficients, extrinsic to the rate constant, V_{max}/K_m, of the enzyme reaction) indicates that a linear free energy relationship (LFER) is observed. Enzyme kinetics in organic media, therefore, can be studied and modeled using physical–organic techniques. For

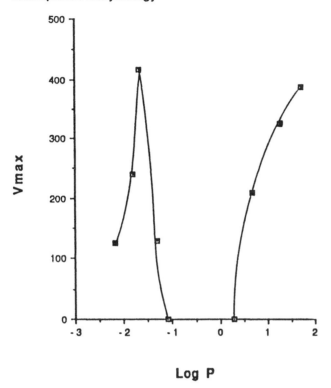

Figure 5 Effect of solvent hydrophobicity on catalytic turnover of horseradish peroxidase. The units of V_{max} are micromoles per milligram of enzyme per second with p-cresol as substrate.

example, peroxidase catalysis in organic media fits the following kinetic description:

$$V_{max}/K_m = 680 \exp\left\{- [0.69(\log P) + 1.50](\pi)\right\} \tag{1}$$

This empirical expression enables peroxidase catalysis in organic media to be predicted solely based on the substrate and solvent employed. Such a "predictive" model is crucial for the optimal design of enzymes in organic solvents.

In addition to providing the framework for modeling reaction kinetics, linear free energy relationships have proved valuable in probing the transition state structure of chemical reactions [63], as well as in enzymatic catalysis in aqueous solution [64-66]. LFER now has been extended to probing the transition state structure of enzymes in organic media using Hammett analysis [67]. The Hammett ρ values of transesterification (in a variety of organic solvents) or hydrolysis in water catalyzed by subtilisin Carlsberg (a serine protease from *Bacillus licheniformis*) are nearly identical, suggesting that the microenvironment of subtilisin's active site and the transition state of the enzymatic reaction in organic media are unchanged from the case of aqueous solutions. This can happen only if the active site is "shielded" from the bulk organic solvent. Once the substrate has reacted with the enzyme, catalysis in organic and aqueous solvents is identical. This hypothesis is supported by structural studies. α-Lytic protease from *Lysobacter enzymogenes* retains the catalytically active triad, typical of a serine protease, in anhydrous organic solvents that support catalytic activity (e.g., acetone and octane): The data are obtained by solid state, magic angle spinning ^{15}N NMR spectroscopy [68]. In dimethyl sulfoxide (DMSO), however, the catalytic triad is significantly disrupted. It is important to note that DMSO fails to support catalytic activity and is capable of disrupting the secondary and tertiary structure of enzymes [32].

Information provided in this section suggests that catalytic activity in organic solvents is a result of the enzyme retaining its native structural configuration. The effect of the organic solvents is to alter the affinity of the substrate with the enzyme's active site. This is not to say, however, that the solvent cannot affect the nature of enzyme catalysis. Both substrate and enzyme stereoselectivity can be altered in organic media solely as a result of changes in the hydrophobicity of the reaction medium.

Substrate and Stereoselectivity Alterations
in Nonaqueous Media

As previously stated, the organic solvent can alter the binding of substrate to enzyme, thereby often causing profound changes in observed enzyme kinetics. One would expect an alteration in the substrate specificity of enzymes in organic solvents as compared to water, and one would expect this alteration to occur in addition to that caused by enzyme rigidity in organic media. This has, indeed, been shown to be the case [16]. Chymotrypsin, subtilisin, and pig liver carboxylesterase contain hydrophobic active sites. In aqueous solutions, nonpolar amino acid derivatives (e.g., esters of phenylalanine and tyrosine) favorably partition from the bulk reaction medium into the enzymes' active sites. This causes an increase in enzymic catalytic efficiency; enzymatic hydrolysis of a phenylalanine derivative is more than 4 orders of magnitude higher than for a hydrophilic serine derivative. In octane, however, hydrophobic partitioning is severely restricted and hydrophilic substrates are favored; the serine derivative is nearly 20-fold more reactive than the phenylalanine derivative.

This pattern also emerged when competitive inhibition was studied with chymotrypsin [19]. In water, naphthalene is an eightfold more potent inhibitor than the hydrophilic 1-naphthoic acid. In octane, the specificity of inhibition reverses, and the acid becomes a 270-fold stronger inhibitor than the naphthalene. Again, these findings can be explained by evaluating the propensity of nonpolar substrates to partition from the bulk aqueous solution to a hydrophobic active site of an enzyme. In octane, this driving force is diminished. It might be expected that some substrates may be more reactive in solvents than in water. A hydrophilic substrate in organic media, provided it remains soluble, would favorably partition into an enzyme's active site. This would lower the apparent K_m of the substrate and increase catalytic efficiency. A case in point is the α-lytic protease catalyzed transesterification of N-Ac-L-Ala-chloroethyl ester with ethanol in octane, which is 80% as efficient as the hydrolysis of the same ester in water [68].

Solvent polarity has been shown to affect enzymic enantioselectivity. Jones and Mehes showed that in a variety of water-miscible organic solvents at concentrations below 40% (v/v), the enantioselectivity of chymotrypsin was diminished by up to 90% [69]. More recently, in a reaction of potential commercial significance, subtilisin was unable to distinguish between L- and D-amino acids and effectively catalyzed the incorporation of D-amino acids into peptides [70].

What causes this behavior? Clearly, the medium exerts a profound kinetic effect on the reactivity of an enzyme. Enantioselectivity depends on the ratio of catalytic efficiencies of the enzyme with both isomers of the substrate (e.g., $(k_{cat}/K_m)_L/(k_{cat}/K_m)_D$). A "relaxation" of enantioselectivity results when the ratio approaches unity. Sakurai and coworkers have found that an increase in medium hydrophobicity causes a reduction in the enantioselectivity of proteolytic enzymes [71]. In fact, an LFER is obtained that correlates the observed enantioselectivity to solvent hydrophobicity (Fig. 6). Changes in enantioselectivity up to 4 orders of magnitude were observed. It has been suggested that this behavior is due to the substrate's ability to displace water from the hydrophobic active site of the serine proteases in water and in organic media [71]. In aqueous media, water displacement from the active site to the bulk solvent is favored; the more hydrophobic the medium, the less favorable the expected displacement of water from the enzymic active site. If water cannot be displaced, L-isomers cannot bind properly, and the k_{cat}/K_m of the substrates is lowered. D-Isomers must bind differently to the enzyme and are less affected by the ability of water to be displaced.

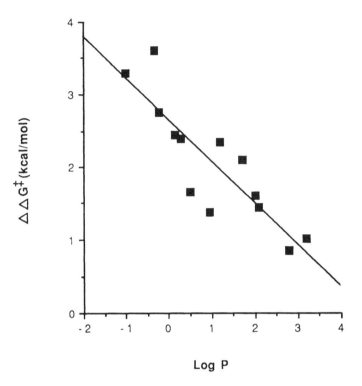

Figure 6 The dependence of the enantioselectivity of subtilisin Carlberg (as represented by the change in the free energy of binding of the L- vs. D-isomers) on solvent hydrophobicity. $\Delta\Delta G^{\ddagger} = -RT \ln[(k_{cat}/K_m)_L/(k_{cat}/K_n)_D]$. The reaction studied was the transesterification of N-AC-Ala chloroethyl ester with 1-propanol. (From Ref. 71.)

Choice of Solvent

Several factors must be taken into account when choosing an appropriate solvent for a given enzymatic reaction—first and foremost, the compatibility of the solvent with the reaction of interest. For example, sugars are soluble only in hydrophilic, water-miscible solvents such as dimethylformamide, formamide, or pyridine. Hydrophobic solvents are inappropriate media for reactions involving sugars because insoluble substrates and insoluble enzymes are unable to interact. Products also must be compatible with the solvent. Polar products tend to remain in the vicinity of the enzyme and can cause product inhibition or undergo unwanted side reactions. This has been observed with tyrosinase-catalyzed oxidation of phenols in hexane [20]. Polar quinone products are insoluble in hexane and undergo undesired polymerization around the enzyme particles, which fouls the enzyme and reduces catalytic activity. In the more polar chloroform, quinones are soluble and the problem is avoided. Also, the solvent must be inert to the reaction of interest. Transesterification reactions involve a nucleophilic attack by an alcohol on an ester to produce a second ester. If the solvent is also an ester, then high conversions of an unwanted solvent-based ester will form at the expense of the desired product. Similar results will occur if the solvent is an alcohol. In some cases, however, an ester or alcoholic solvent is desirable if it acts either as the acyl donor or acceptor (nucleophile) in the reaction; the high concentration of solvent in the reaction mixture can drive equilibrium-controlled reactions to completion.

Additional factors that may influence the choice of solvent include solvent density and viscosity, surface tension, toxicity, flammability, waste disposal, ease of recycling and/or regeneration, and cost.

In general, the more hydrophobic the organic solvent, the more active an enzyme is likely to be when placed in it (see Table 2 for a representative example). As stated, this is because hydrophobic solvents cannot easily strip away the essential water associated with an enzyme. Attempts to correlate activity with solvent hydrophobicity generally have focused on the log P model (defined above) advanced by Laane and coworkers [61,72]. Solvents with log P values above 4 (e.g., decanol, hexadecane, diphthalate esters), nearly always support enzymatic activity. Unfortunately, most compounds of interest, both for preparative synthesis and fundamental studies, are insoluble in such extremely hydrophobic media.

The danger in the log P model is the tendency to neglect that it is highly empirical; catalytic activity is correlated to an extrinsic measure of solvent hydrophobicity. The log P model does not provide a kinetic or mechanistic basis of solvent compatibility to enzymatic catalysis. In addition, many correlations of catalytic rates with solvent hydrophobicity fail to take into account the diminished substrate binding to a hydrophobic active site of an enzyme in hydrophobic media. Rate measurements are often taken at a substrate concentration similar to that used in water. Increases in substrate K_m will be manifested in lower observed rates of catalysis. Yet, catalytic turnover is not always reduced in organic media [57,59], resulting in suboptimal use of a given enzyme in the organic solvent. The advantages of the log P model are simply that it provides a framework for choosing solvents that are expected to support catalytic activity, and that log P is a quantitative, extrinsic solvent parameter that can be used in linear free energy relationships.

Biocatalyst Preparation

The insolubility of enzymes in monophasic organic solvents has a controlling influence on the kinetics of enzymatic catalysis. Intraparticle and external diffusional resistances not normally encountered with freely soluble enzymes must be taken into account when using enzymes in organic media, and the intrinsic kinetics of enzymatic catalysis in organic solvents are often masked. This problem is most severe with highly active enzymes.

Conventional approaches to relieving diffusional limitations in heterogeneous chemical catalytic systems have been used in enzymic reactions in organic solvents. One particularly attractive method is to spread the enzyme onto materials with large surface-area-to-volume ratios such as nonporous glass beads [20,38]. Because enzymes are insoluble in organic solvents, the bound enzyme will have no propensity to desorb from the glass surface, and facile immobilization is feasible without the cumbersome covalent attachment or entrapment procedures required of enzymes in aqueous media. Rate enhancements above 2 orders of magnitude are routinely obtained using this procedure with horseradish peroxidase [31].

Adsorption of an enzyme onto the glass beads does not guarantee elimination of intraparticle diffusional limitations. Multiple enzyme layers around the glass bead can be obtained, thereby causing inadvertent diffusional resistances that can be as severe as in insoluble enzyme pellets. This "overcrowding" could be relieved if monolayer enzyme coverage were achieved. As shown in Figure 7, monolayer coverage is attained at an enzyme loading of 0.1 mg per gram of beads at a bead size of 75–150 μm. This value correlates well with the expected surface area coverage of peroxidase as calculated from enzymatic hydration radius geometries [73].

Internal diffusion can be optimized, theoretically, by solubilizing enzymes in organic solvents. Inada and coworkers have solubilized several enzymes including horseradish peroxidase, catalase, lipoprotein lipase, and chymotrypsin by modifying enzymes' free amino groups with a 5000 dalton polyethylene glycol (PEG) derivative [74–76]. The enzymes are soluble and active in a variety of polar and nonpolar solvents. A small amount of water is required for activity, and it is not clear how enzyme hydration is maintained. (One possibility is that the amphiphilic PEG chains form a protective shell around a solubilized enzyme and entrap water, much as in reverse micelles[77].) Chemical acylation of nucleophilic groups on an enzyme may also lead to soluble preparations in organic solvents. Horseradish peroxidase, acylated with palmitoyl chloride, was soluble in dioxane, containing 5% (v/v) water. It also retained up to one-third the initial enzyme activity [78]. The process is far simpler than PEG modification. From a practical standpoint, solubilized enzymes in or-

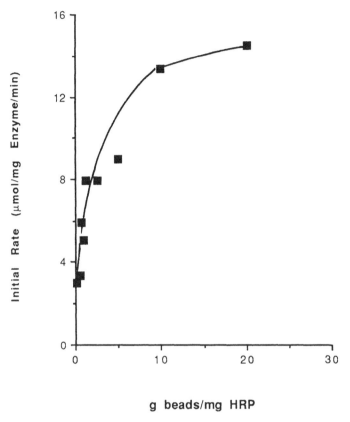

Figure 7   Effect of peroxidase loading onto glass beads (75–150 μm). (From Ref. 38.)

ganic media cannot be used in a continuous fashion, and this discourages their use in preparative bioconversions.

Enzyme kinetics using immobilized enzyme preparations may be affected by the support, even when diffusional limitations are eliminated. Hydrophilic supports, while providing a stabilizing effect for enzymes, lower the solubility of hydrophobic substrates in the vicinity of the enzyme, thereby increasing the apparent K_m of the reaction and reducing catalytic efficiency [79]. furthermore, hydrophilic supports may compete with the enzyme for available water in the reaction system. The resulting insufficiently hydrated enzymes show reduced catalytic activity [80]. Hydrophobic supports will not compete with the enzyme for water but can cause inactivation [81].

APPLICATIONS OF ENZYMES IN ORGANIC MEDIA

The study of organic solvent effects on enzymatic reactions began in the late nineteenth century. Chiltenden and Mendel observed that trypsin-catalyzed proteolysis could be inhibited by adding 20% (v/v) alcohol, the inhibitory effect becoming more pronounced as the molecular weight of the alcohol increases [82]. Twenty years later, considerable activity in crude preparations of urease, β-glucosidase, β-galactosidase, invertase, papain, peroxidase, and catalase were observed in 90% (v/v) solutions of alcohols [83]. These enzymes were insoluble suspensions in alcoholic solvents and represented an early, yet successful application of immobilized enzymes.

The main purpose of adding organic solvents to aqueous solutions of enzymes was to enhance the solubility of nonpolar substrates. The use of enzymes as biocatalysts in monophasic organic solutions dates back to the mid-1960s with the work of Dastoli and Price [84]. They suspended moist preparations of chymotrypsin and xanthine oxidase in a number of nonpolar solvents and observed small, yet measurable amounts of catalytic activity. Siegel and Roberts obtained similar results for horseradish peroxidase and catalase suspensions in a variety of water-miscible and -immiscible solvents [85].

The applications of enzymes in monophasic organic media described in this chapter are categorized by conventional enzyme classifications. These include oxidoreductases, transferases, hydrolases, and isomerases. These applications include reactions important in organic synthesis and polymer production, and as analytical reagents. Organic media represent organic solvents, supercritical fluids, and gas phase organics.

Oxidoreductases

A great deal of research has been carried out using oxidoreductases in nonaqueous media, particularly with oxidases and dehydrogenases. In most cases, oxidoreductase substrates of commercial interest are poorly soluble in water, and significant solubility enhancements are observed in organic systems.

Steroid Oxidases and Dehydrogenases

Steroids and their derivatives are useful as drugs, hormones, and animal feed supplements [86]. Due to low steroid solubilities in aqueous solutions, process yields have been poor. Monophasic organic solvents systems have, under rigidly controlled conditions, eliminated this problem. The simplest oxidations are those that are cofactor independent. Purified cholesterol oxidase, adsorbed to glass beads, was shown to be highly active in toluene, oxidizing cholesterol to cholestenone with high efficiency [21]. Preparative synthesis ($\approx$ 160 g/L) of cholestenone in carbon tetrachloride has been carried out using wet cells of a *Nocardia* species [87]. Cholesterol oxidase has been shown to be highly active in supercritical CO_2 [88.89]. Cosolvents such as ethanol or *tert*-butanol significantly enhanced the reactivity of the enzyme. Because of the high solubility of cholesterol in supercritical CO_2 ($\approx$ 50 times more soluble than water), higher turnover numbers of up to 100-fold were obtained, compared to aqueous solutions. Interestingly, the use of cholesterol oxidase in organic solvents allowed a mechanistic evaluation of the enzymatic reaction that could not be surmised in aqueous solutions, namely that cholesterol oxidase is more active on cholesterol aggregates than on cholesterol monomers [88]. Such a finding indicates that cholesterol oxidase is stimulated by a hydrophobic surface, such as that provided by a cholesterol aggregate. Cholesterol oxidase is an ideal steroid-transforming catalyst because no cofactors are required.

The use of whole cells is nearly always mandated for steroid transformations requiring a NAD(P)(H) cofactor, because cellular cofactor regeneration pathways can be utilized even when the cell is not growing, as in an organic solvent. Such applications of whole cells have widened the scope of steroid oxidation to include the steroid dehydrogenases. Oxidation of 3β-hydroxy Δ-5-steroids to the 3-keto derivatives have been carried out in a 1:1 mixture of hexane–benzene using whole cells of *Nocardia rhodocrous* [90]. Pregnenolene dehydrogenation to progesterone and β-estradiol oxidation to estrone could also be carried out using *N. rhodocrous* cells in organic media [91].

The preparation of the biocatalyst has been shown to be a vital parameter for selective oxidations. Modifying the hydrophobicity of a gel matrix entrapping the whole cells affects the reactivity of the cell toward various steroid substrates [90]. Hydrophilic polyurethane gels lower the activity of cholesterol oxidation catalyzed by *N. rhodocrous* cells while enhancing the activity of dehydroepiandrosterone (DHEA) oxidation, whereas the reverse is true with hydrophobic polyurethane gels. Cholesterol, being less polar than DHEA, is less capable (relative to DHEA) of partitioning through the hydrophilic polyurethane. The lower cholesterol concentration in the vicinity of the entrapped cells causes a reduced observed reactivity. Similar results were observed when a hydrophilic electron acceptor, phenazine methosulfate (PMS), was required; hydrophilic gels enhanced the partitioning of the polar PMS into the vicinity of the biocatalyst, thereby enhancing reactions that require an external electron acceptor [92]. A more complete description of steroid biotransformations is provided in Chapter 5 of this book.

Alcohol Dehydrogenases

Alcohol dehydrogenases catalyze the oxidation of alcohols and aldehydes and the reduction of carbonyl compounds. The reactions depend strictly on a cofactor such as NAD(H). Because the cofactor must be recycled, alcohol dehydrogenases used in commercial applications carry out a secondary reaction, either oxidation of an alcohol or reduction of an aldehyde, which is mechanistically opposite to the reaction of interest. The cofactor recycling in aqueous solutions is rapid, yet cofactor stability is poor; recycling numbers for NAD^+ rarely exceed 2000 [93].

Grunwald and coworkers studied the stereospecific oxidation of alcohols to optically active ketones using horse liver alcohol dehydrogenase (HLADH) in isopropyl ether [94]. The NADH produced during the reaction is oxidized by concurrent reduction of isobutyraldehyde to isobutanol. HLADH is adsorbed to nonporous glass beads simultaneously with NAD^+. Because neither cofactor nor enzyme is soluble in organic solvents, enough water is added to the reaction mixture ($\approx 1.5\%$ v/v) to provide mobility of the nicotinamide cofactor on the glass bead surface, hence to allow interaction with the enzyme. As a result, cofactor recycling is possible and turnover numbers exceeding 2 million are obtained. The major drawback is the long reaction time compared to chemical synthesis; yet this is compensated for by high levels of enantioselectivity (enantiomeric excess 95-100%). The asymmetric reduction of racemic aldehydes and ketones can also be performed with HLADH [94]. Regeneration of NADH is accomplished by concurrent oxidation of ethanol.

Deetz and Rozzell have studied the kinetics of HLADH and yeast alcohol dehydrogenase (YADH) in various organic solvents [95]. Hydrophobic, water-immiscible solvents sustain catalytic activity, but cofactor mobility is severely restricted. Butyl acetate, a solvent of moderate polarity and containing 1% v/v water, was found to be an excellent solvent. Product inhibition, often severe in aqueous solutions, is up to 3 orders of magnitude less severe in organic media. Furthermore, saturation of enzyme with NAD^+ occurs with lower concentrations than in water; hence less of the expensive cofactor is required. Alcohol dehydrogenase has been used in the gas phase; continuous oxidation of gaseous butanol to butynaldehyde was achieved over 25 hours [96].

Peroxidases

Peroxidases catalyze the one-electron oxidation of phenols and aromatic amines in the presence of hydrogen peroxide. Their most prevalent natural reaction is the formation of lignin in plants [97]. The enzyme is highly active in organic solvents, with turnover numbers often enhanced in nonaqueous media over that in water [38,59]. Horseradish peroxidase has been used in a number of reactions of commercial interest in monophasic organic media, including phenolic polymerizations and lignin depolymerization, and in analytical sensing devices.

Phenolic polymerization is the core process in the synthesis of phenol–formaldehyde resins. These materials are used routinely as laminates, adhesives, bonding agents, and photographic developer resins, in particle board, as soil conditioning agents, and as foams and insulators [98]. The industry faces a grave problem as a result of the presence of formaldehyde in polymers, both in resin manufacture and in the end products. Formaldehyde recently has been identified as a severe health and environmental hazard [99], and alternative methods to produce phenolic resins are needed. One such alternative is to use peroxidase to catalyze the polymerization of phenols in the absence of formaldehyde.

The application of peroxidase for phenolic polymerization has been attempted in aqueous solutions, yet only low molecular weight coupling products are formed [100]. High molecular weight polymers cannot be synthesized because of the poor solubility of the growing polymer chains in water. Poor substrate solubility also lowers the productivity of phenolic polymerization in water. Dordick and coworkers have examined the polymerization of phenols in nonaqueous media in order to overcome poor substrate and polymer solubilities and low polymer molecular weights [101]. When p-phenylphenol, a hydrophobic phenol of great commercial interest in the synthesis of oil-soluble phenol–formaldehyde resins [102], was used, peroxidase catalyzed polymerization in dioxane (containing 15% aqueous buffer) yielding molecular weights in excess of 25,000 daltons, nearly 50-fold higher than that obtained in aqueous solutions. Furthermore, the productivity of poly(p-phenylphenol) in dioxane is far higher than in water due to the high solubility of the phenol in dioxane. Using 1 M initial concentration of the phenol in 85% dioxane, 150 g/L poly(p-phenylphenol) was produced. This compares favorably with the maximum solubility of p-phenylphenol in water of about 1.5 mM. In addition to increased size of the polyphenols, peroxidase catalysis in dioxane can be controlled selectively [101]. Polymer size is found to be highly sensitive to the water content in dioxane; below 30% v/v water, relatively small polymers are produced, whereas in 15% v/v water, molecular weights in excess of 25,000 daltons are produced. Specific molecular weights can be obtained by varying the water concentration in dioxane (Fig. 8).

The polymerization of phenols catalyzed by peroxidase in organic media is a general phenomenon. A wide variety of electron donors are capable of acting as substrates (Table 3), as well as a number of water-miscible solvents capable of supporting the polymerization reaction. This reaction has been commercialized as a phenol–formaldehyde replacement [103].

In an opposite reaction to phenol polymerization, peroxidase has shown limited activity on lignin [78]. Depolymerization to low molecular weight oligomeric products (MW $\approx$ 500) was carried out in dioxane supplemented

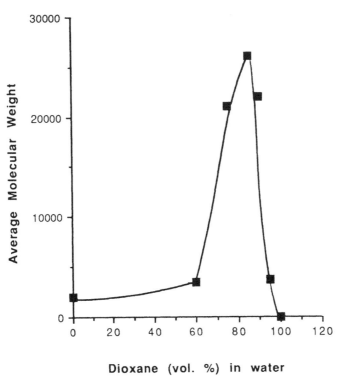

Figure 8   Peroxidase-catalyzed polymerization of *p*-phenylphenol in dioxane–water mixtures: effect of poly(*p*-phenylphenol) size on dioxane concentration. Molecular weight determined via LH-60 Sephadex gel permeation chromatography and capillary viscometry (75–90% dioxane v/v, only). (From Ref. 101.)

with 5% aqueous buffer. Synthetic, natural (milled wood), and processed (kraft) lignins were used as substrates and up to one-third of the high molecular weight lignins were partially depolymerized. Interestingly, as the water content of the dioxane was increased, depolymerization activity decreased in favor of polymerization [31].

Peroxidases have also been employed as analytical reagents in monophasic organic media [21,104]. A common approach in conventional aqueous-based enzymatic analysis is to couple peroxidase with a selective oxidase for quantification of an analyte. In a clinically relevant example, cholesterol, extracted into toluene from an aqueous solution, can be efficiently measured using a bienzymic system of cholesterol oxidase and horseradish peroxidase (Equa-

Table 3 Spectrum of Compounds That Are Polymerized by Horseradish Peroxidase in Dioxane[a]

Compound	Average molecular weight (daltons)
Phenol	1400
p-Methoxyphenol	2000
p-Cresol	1900
p-Chlorophenol	600
2,6-Dimethylphenol	500
4,4'-Biphenol	400
Aniline	1700
1–Naphthol	Very high[b]
2–Naphthol	2000
p-tert-Butylphenol	1900
p-Phenylphenol	26,000[c]

[a]Reaction conditions: 20 mM substrate, 20 mM H_2O_2 (added slowly throughout the reaction), dioxane (containing 15% aqueous acetate buffer, 10 mM, pH 5.0). Unless otherwise stated, molecular weights are weight averages from LH-20 Sephadex chromatography.
[b]Insoluble product that was unable to be analyzed by GPC.
[c]Molecular weight obtained via viscometry.
Source: Ref. 101.

tions 2 and 3). There was excellent agreement between cholesterol measured and that actually added to an independent solution in toluene [21].

$$\text{cholesterol} + O_2 \xrightarrow{\text{cholesterol oxidase}} \text{cholestenone} + H_2O_2 \tag{2}$$

$$H_2O_2 + p\text{-anisidine} \xrightarrow{\text{horseradish peroxidase}} \text{colored product} + 2H_2O \tag{3}$$

The production of color during peroxidase catalysis on selected chromogenic substrates also has been used in the development of a sensor that can measure temperature changes [104]. While peroxidase functions well in liquid organic solvents, it is unable to do so in solid media, such as solid paraffins; the rate of p-anisidine oxidation catalyzed by peroxidase is more than 5 million–fold slower in solid hexadecane (at 4°C) than in liquid hexadecane (at 25°C). The diminished catalytic activity is surely the result of diffusional barriers inherent in solid media. A prototype device was developed that takes advantage of the inactivity of peroxidase in a solid solvent for accurate, reliable, and irreversible (tamper-proof) temperature abuse detection (increase of temperature beyond prescribed limits) [105]. As the temperature rises above the melting point of the paraffinic solvent, the peroxidase quickly catalyzes the oxidation of *p*-anisidine and color is produced. By mixing various hydrocarbons, the melting point of the solvents can be varied. This abuse sensor may find particular applications in the transportation of perishables such as frozen or chilled foods, pharmaceuticals, and other heat–sensitive materials.

General Oxidases and Oxygenases

A variety of oxidases and oxygenases of potential commercial importance have been studied in monophasic organic media, including epoxidases, hydroxylases, phenol oxidases, and alcohol oxidases.

Brink and Tramper carried out the microbial production of propylene oxide in *n*-hexadecane with *Mycobacterium* sp. cells entrapped in a calcium alginate gel [106]. Unfortunately, the solubility of the gaseous propylene was poor in the microaqueous environment within the gel matrix and reaction rates were poor. Extension of this reaction to a hydrophobic polyurethane gel led to biocatalyst inactivation.

Few reactions involving aliphatic hydroxylation have been performed in monophasic organic media. Takazawa and coworkers observed that tetradecane hydroxylation to tetradecanol could be catalyzed by whole cells of *Corynebacterium equii* in isooctane, hexane, and toluene [107]. Preparative synthesis was limited by the further oxidation of the alcohol to the fatty acid.

Regioselective hydroxylations have been performed using a combined enzymochemical approach with polyphenol oxidase (tyrosinase) in chloroform [20]. The enzyme efficiently oxidizes *p*-cresol to the orthoquinone. In chloroform, the quinone is highly stable and can be reduced in the presence of ascorbic acid to 4-methylcatechol, thereby affording a regioselective hydroxylation of the phenol. Gram-scale conversions were carried out.

Polyphenol oxidase has also been used in supercritical fluids such as CO_2 and fluoroform [108]. In both static and continuous systems at 35°C and 5000 psi, *p*-cresol was converted to 4-methylcatechol, which subsequently underwent polymerization. However, the enzyme was quickly inactivated. This phenomenon is similar to the oxidation of *p*-cresol in hexane and is due to the inability of the quinone product to partition away from the enzyme microenvironment and into the supercritical solvent. This is not unexpected, based on the hydrophobic similarity of supercritical CO_2 and hexane [109].

Laccase, an oxidase similar to peroxidase in substrate specificity but one that does not require H_2O_2 for activity, has been studied in organic media [110,111]. The enzyme obeyed Michaelis–Menten kinetics in 17 solvents either saturated with water or in the presence of 3.5% (v/v) water (water immiscible solvents only). Reaction rates up to 10% of that in water were observed. Laccase was also employed in dioxane solutions (containing 30% water) in the oxidation of water-insoluble lignins. Gel permeation chromatographic (GPC) analysis revealed an increase in the molecular weight of the lignin polymers in the organic solvent, similar to that observed with horseradish peroxidase in 70% dioxane [78].

Alcohol oxidase from *Pichia pastoris* has been used in organic media for the production of hydrogen peroxide using methanol and ethanol as alcoholic substrates in kerosene–tributylphosphate mixtures [112]. A relationship existed between the concentration of H_2O_2 produced and its solubility in the solvent. Simultaneous purification and concentration of the H_2O_2 could be carried out by extraction into water, thereby eliminating the necessity of distilling large amounts of dilute H_2O_2 streams. Alcohol oxidase has also been used in the gas phase for the oxidation of ethanol [113]. The enzyme was deposited onto DEAE-cellulose particles and catalase was codeposited to prevent H_2O_2-based inactivation of the enzyme. Such a system has been converted into an enzyme sensor for the gas phase level of ethanol; this type of application is exemplified by the breathalyzer [114].

Transferases

Very few transferases have been used in organic solvents. This is not surprising, because these enzymes generally have polar substrates, including amino and nucleic acid derivatives, carbohydrates, and sulfate and phosphate derivatives. There are some selective advantages in using transferases in monophasic organic media, however, particularly when competing hydrolytic reactions are possible. Such is the case with levansucrase. This enzyme catalyzes the synthesis from sucrose or raffinose of the fructose polymer levan [115]. In aqueous solutions, the enzyme also catalyzes the hydrolysis of both levan and sucrose, thereby lowering the yield of levan in the final product. In acetonitrile (containing 30% water), levansucrase exhibits transferase activity only, while catalytic turnover increases by nearly fivefold.

Hydrolases

In aqueous solutions, enzymes such as lipases, esterases, proteases, and carbohydrases catalyze hydrolytic reactions. These enzymes are, for the most part, mechanistically well understood, highly stable, cofactor independent, and commercially available. These are attributes that make them well suited for commercial applications both in aqueous solutions and in organic media. They also carry out hydrolysis, a reaction with limited commercial appeal. In organic media, however, these enzymes catalyze a variety of reactions including esterification, transesterification, thiotransesterification, interesterification (acyl exchange), lactone synthesis, aminolysis, and oximolysis [3,41]. All these reactions are possible only in low water environments. Because hydrolytic enzymes are mechanistically similar, specific hydrolases will not be grouped into separate classifications; rather, general categories of organic synthesis will be examined.

Ester Synthesis: General

Lipases, both as free enzyme powders and in whole cells, have been used to catalyze esterifications in nearly anhydrous solvents. Bell and coworkers performed lipase-catalyzed fatty acid acylations by suspending cells of *Rhizopus arrhizus* in octanol containing palmitic or oleic acid [116]. Yields of the octyl esters in excess of 70% were obtained both in batch and continuous systems [117]. The lipase remained active in the solvents for a month with no loss in operational stability. Similar synthetic reactions have generated much interest in the production of geranyl butyrate [118], menthyl acetate [119], waxes and emulsifiers [120,121], and stereo- and regioselective preparative acylations for intermediates in chemical synthesis [4,122,123]. In some cases, conventional esterification has given way to transesterification, whereby a good leaving group on the acyl donor is used and high yields of ester synthesis can be obtained [41].

Lipase catalysis has been extended to supercritical fluids. Van Eijs and coworkers [124] used a commercially available enzyme in supercritical CO_2 to catalyze the transesterification of isoamyl alcohol with ethyl acetate, yielding isoamyl acetate with a maximal productivity of 120 g of isoamyl acetate per kilogram of enzyme per hour.

Enzymes other than lipases have also been used to catalyze ester synthesis in organic solvents. Chymotrypsin has been used to synthesize aromatic amino acid esters with yields approaching 95%, even in the presence of 2.5–10% (v/v) water [125,126]. A similar esterification was performed using chymotrypsin immobilized to a porous polyethylene terephthalate support suspended in ethanol–chloroform mixtures [127]. Once again, yields in excess of 90% were obtained. The plant protease papain has been used to synthesize a variety of *N*-substituted amino acids. Unlike chymotrypsin, papain has a wide substrate specificity and can be used to esterify at least 15 of the 20 naturally occurring amino acids [128,129]. Tannase, immobilized to porous silica, has been used to synthesize esters of gallic acid [130]. Of particular interest is the synthesis of propyl gallate, an effective antioxidant in the food industry.

Like lipases, proteases have been employed in supercritical fluids. Subtilisin catalyzed the transesterification between *N*-Ac-L-Phe-chloroethyl ester and ethanol [131]. The enzyme was highly stable at 45°C and 150 bar, and yields approaching 100% in less than one hour were obtained, compared to yields below 30% in 2 hours in a hydrophobic organic solvent such as benzene.

Regioselective Acylation of Glycols and Polyols

The specificity of enzymes in organic solvents is clearly evident in the regiospecific acylations of multifunctional molecules. These compounds range from the relatively simple aliphatic diols and aminodiols to complex oligosaccharides. Chemical syntheses universally lack regioselectivity, giving mixtures of products [132,133] and requiring expensive protection and deprotection steps for regiospecific synthesis [134,135].

Porcine pancreatic lipase has been found to be an excellent regiospecific catalyst for transesterification of primary hydroxyl groups [136]. In the presence of multiple primary and secondary hydroxyls, the enzyme acylated only the primary groups. An advantage of this type of reaction is the ability to use an ester (such as ethyl acetate or ethyl butyrate) as both acyl donor and solvent, providing for greater equilibrium yields of the desired product.

Lipases and proteases are also capable of acylating specific classes of nucelophilic groups. The lipase from *Aspergillus niger* in *tert*-amyl alcohol catalyzed the acylation of the alcoholic moiety of 6-amino-1-hexanol 37 times faster than that of the amino group. Similar results were observed with lipases from porcine pancreas and *Pseudomonas* sp. [137]. Chemical acylation favors the more nucleophilic amino group [138]. Subtilisin, conversely, catalyzed acylation of the amino group 50-fold faster than the hydroxyl group [137]. This finding could be useful in preparing selectively blocked multifunctional intermediates in organic synthesis in addition to forming selective peptides containing hydroxyl groups (e.g., those containing serine or threonine).

Lipases and proteases, together, have been used in the regiospecific oxidation of steroids and in the absence of oxidoreductases [139]. Preparative-scale esterification in anhydrous acetone of dihydroxy 5-androstan–3β, 17β-diol with trifluoroethylbutyrate with the lipase from *Chromobacterium viscosum* yielded the 3β-monobutyryl ester as the sole product. Once the steroid has been resiospecifically esterified, chemical oxidation of the free hydroxyl group can be performed, followed by chemical hydrolysis of the ester. This method, highlighted in Figure 9, affords regiospecific and preparative-scale steroid oxidations.

In addition to relatively simple aliphatic diols, amino diols, and steroid diols, regioselective acylation of sugars has been carried out. Porcine pancreatic lipase suspended in pyridine (one of the few solvents capable of dissolving sugars) and in the presence of activated trichloroethyl carboxylate esters ranging in size from acetate to laurate, catalyzed the preparative monoacylation of glucose, galactose, and mannose, and the diacyltion of fructose [17]. Gram

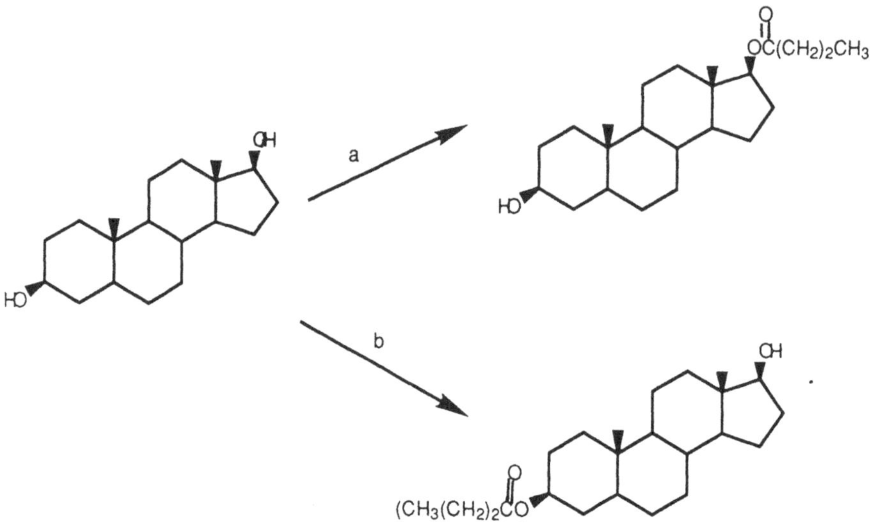

Figure 9 Regioselective acylation of bifunctional steroids with trifluoroethyl butyrate: (a) subtilisin catalysis and (b) *Chromobacterium viscosum* catalysis. (From Ref. 139.)

amounts of primary acylated sugars were produced. Regioselective synthesis of secondary monosaccharide esters were obtained by a similar approach using lipases from porcine pancreas, *Chromobacterium viscosum*, and *Aspergillus niger*, although protection of the primary hydroxyls was required prior to secondary acylations [18]. True esterifications of simple glucosides (e.g., methyl through *n*-butyl glycosides) with octanoic through octadecanoic acids was carried out with lipases from *Candida antartica, Mucor miehei,* and *Humicola* sp. [140].

Replacing the trichloroethyl carboxylate with a triglyceride allowed for the facile synthesis of sugar-based surfactants [141]. For example, in the presence of natural plant and animal oils in pyridine, porcine lipase catalyzed the monoacylation of sorbitol in a single primary position with moderate yields. The enzyme is relatively nonspecific with respect to fatty acid composition or position on the triglyceride. The sorbitol monoesters were found to possess excellent surfactant properties, including the ability to lower interfacial tension and stabilize emulsions. Sorbitol could be replaced with other sugar alcohols such as xylitol, ribitol, arabitol, and mannitol. A significant result of this reaction is the low yields of, di- and multiacylated sugars. Such products would surely reduce the surfactant quality of the product. As another substitute for the trichloroethyl carboxylate, isopropenyl and vinyl acetates and propionates have been used to acylate sugars, glycols, polyols, and organometallics [142]. The leaving group is the enolate of acetone or acetaldehyde, which favorably tautomerizes to acetone or acetaldehyde, respectively, thereby both eliminating the possibility of leaving group attack onto the newly formed ester and simplifying product recovery.

Di- and trisaccharides, such as sucrose, lactose, trehalose, and raffinose, have been acylated by a number of lipases in pyridine [143]. In the case of sucrose, high yields of sucrose 1,4'-diacetate was obtained in the presence of isopropenyl acetate using a lipase from species of *Pseudomonas*. Subtilisin has been used in dimethylformamide to selectively acylate sucrose, maltose, lactose, cellobiose, and oligosaccharides, as well as derivatives of nucleotides and amino acids [144.145]. In all cases, primary hydroxyl groups were acylated.

Interesterifications

A commercially important type of esterification is acyl exchange, or interesterification. Interesterifications have been used primarily in the oil and fat industry to modify the composition and physical properties of triglycerides. For example, Matsuo and coworkers used a 1,3-positionally specific lipase from *Rhizopus niveus* to catalyze the interesterification of 1,3-dipalmitoyl-2-oleoyl glycerol—which is a major triglyceride of palm oil—with either stearic acid or tristearine, to give mixtures of 1-palmitoyl-2-oleoyl-3-stearoyl glycerol and 1,3-distearoyl-2-oleoyl glycerol, both major components of cocoa butter (Fig. 10) [146]. Hence, a higher value triglyceride can be prepared from low value starting materials. The modification of triglycerides has also been used in the margarine industry to produce lower melting temperature, hence more spreadable, margarines [147]. By exchanging saturated acyl moieties in existing margarines with unsaturated fatty acids, melting points have been lowered by as much as 30°C.

Interesterifications require that sufficient water be available for hydrolysis of fatty acids in a triglyceride, yet not enough water to prevent esterification

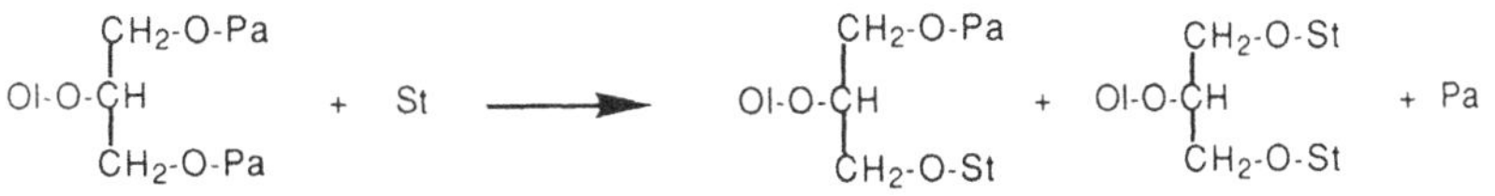

Figure 10 Interesterification of POP with stearic acid to give POSt and StOSt catalyzed by *Rhizopus niveus* lipase: Pa, palmitic acid; St, stearic acid; and Ol, pleic acid. (From Ref. 146.)

with a new fatty acid or transesterification with a fatty acid ester. Water control is difficult and is sensitive to the enzyme environment [148]. High water contents in commercial interesterifications often lead to unacceptable product mixtures [149].

Lactone Synthesis

Lactone synthesis in organic media was first noted by Gatfield using a fungal lipase [120,150]. Pentadecanolide was synthesized from 15-hydroxypentadecanoic acid, and γ-butyrolactone was prepared from 4-hydroxybutyric acid. Efficient stereoselective lactone synthesis has been carried out by Gutman and coworkers with γ-hydroxyacids catalyzed by porcine pancreatic lipase in hydrophobic solvents [151]. Similarly, in the presence of racemic methyl 10-hydroxyundecanoate, *Pseudomonas* sp. lipase catalyzed stereoselective lactone formation [152]. A significant amount of macrocyclic lactones was formed as a result of intermolecular transesterification. The length of the hydroxyacid is known to control intra- or intermolecular lactone formation. 10-Hydroxydecanoic acid undergoes macrocyclic lactone formation, while 16-hydroxyhexasecanoic acid undergoes primary monolactone formation [153].

Deliberate intermolecular lactonization has been carried out by Chen and coworkers [154]. Lipases from *Candida cylindracea, Pseudomonas* sp., and porcine pancreas catalyzed the condensation of diols and diacids in isooctane, carbon tetrachloride, and cyclohexane. The reaction obeyed an unusual temperature dependence. At temperatures from 55 to 75°C, yields of the monolactone products were 23–38%. At temperatures below 45°C, yields of the monolactone products were below 4%, while oligomeric esters formed the predominant product. Evidently a higher activation energy was required for lactonization than for polymerization.

Lipase-Catalyzed Polymer Synthesis

The prospect of carrying out stereoselective polyester synthesis has attracted the attention of a number of groups. Ajima and coworkers synthesized poly(β-hydroxyesters) in a variety of solvents catalyzed by an organic-soluble, PEG-modified lipoprotein lipase [155]. The procedure was reported to have led to the synthesis of high molecular weight polyesters, but no firm molecular weight or structural data were given. Stereoselective polycondensations of racemic diesters with achiral diols, and achiral diesters with racemic diols, were performed with lipases from a species of *Chromobacterium* and from *A. niger* in very dry toluene [156]. While only one isomer of each racemate was incorporated into the product, only oligomeric products were obtained. Typically, ratios of diester to diol were 1:4. This led to oligomers terminated in alcohol functionalities and prevented stepwise polymerization to high molecular weight polymers.

High molecular weight polymer has been synthesized using porcine pancreatic lipase in diethyl ether [157]. Condensation of bis(2,2,2-trichloroethyl)($\pm$)-3,4-epoxyadipate with 1,4-butanediol in equimolar concentrations (with respect to the ester isomer recognized by the enzyme), yielded polymers with M_n = 5300 daltons. Furthermore, the product was shown to be optically active, indicating that only one isomer of the adipate derivative was consumed. Oligomeric polycarbonates have been synthesized using lipase from *C. cylindracea* and porcine liver esterase in a variety of water-miscible and -immiscible solvents [158]. Finally, sucrose-based polyesters have been prepared using an alkaline protease in pyridine [41a]. Molecular weights in excess of 13,000 were obtained.

Peptide Synthesis

Peptide synthesis in monophasic organic systems has been performed in solvents with water contents ranging from nearly anhydrous [159] to 50% water [160]. In general, peptide synthesis has been carried out with amino acid derivatives that have appreciable solubility in organic solvents. For example, chymotrypsin immobilized to Sepharose catalyzes the coupling of N-Ac-L-Phe-OMe with alanine and glycine amino acid amides in butanediol with a water content of 10% (v/v) [160]. The yield of dipeptide increased as the water content of the solvent decreased. Similar results were obtained using chymotrypsin immobilized to chitin in 95% acetonitrile for the synthesis of N-Ac-L-Tyr-L-Gly-NH2 [126]. Enzyme stability in these systems, however, was poor. Furthermore, unwanted hydrolysis of the products reduced the yields [161].

Much more stable enzyme preparations and high product yields were possible in water-immiscible solvents with low water content. Thermolysin has been used to synthesize a variety of dipeptides [162,163]. Perhaps one of the most important commercial uses of thermolysin is in the synthesis of the dipeptide sweetener Aspartame [159,164]. Improved yields were obtained in ethyl acetate containing 2% aqueous buffer; the water was shown to be bound entirely to the biocatalyst particles. Thermolysin catalysis in water-immiscible organic solvents can be activated by a variety of cosolvents such as ethylene glycol, glycerol, and formamide [28]. These compounds act as water mimics and enable reactions to be performed in lower water concentrations, resulting in a nearly complete elimination of the water-driven competing hydrolysis reaction and enabling polypeptides to be prepared.

In addition to the use of proteases in organic media, lipases have found promise in peptide synthesis [165,166]. Because lipases do not exhibit amidase activity, peptide hydrolysis is prevented and high yields of peptide synthesis products are possible. Lipases, however, do not offer the rigid substrate specificity available with proteases for amino acid substrates; hence, general use for selective peptide synthesis may be limited.

One solution to the lack of selectivity afforded by lipases in peptide synthesis is to alter the activity of proteases to eliminate amidase activity. Wong and coworkers [167] used a chemically modified chymotrypsin (methylated at His57) that retained esterase activity but not amidase activity. In 50% DMSO solutions (a high enough organic solvent concentration to warrant inclusion in monophasic discussions), a variety of dipeptides were prepared in high yields (see Table 4 for representative examples). Water-miscible solvents also have been shown to affect the ratio of esterase to amidase activities of serine proteases [168]. Above 60% (v/v) acetonitrile, DMF, and dioxane, trypsin retained no amidase activity yet retained nearly full esterase activity. This finding suggested that kinetically controlled peptide synthesis was possible (Table 4) with very little loss due to product hydrolysis.

Preparation of Optical Isomers

The production of optically active compounds is a great challenge to synthetic organic chemists. Hydrolases have been routinely used for the resolution of racemic alcohols and carboxylic acids through the hydrolysis of the corresponding esters [170,171]. In organic media, this approach has been extended to esterifications, transesterifications, and aminolysis. The procedure is straightforward: racemic alcohols or amines are esterified or transesterified with an achiral carboxylic acid or ester, respectively, to produce a chiral ester or amide that can be isolated from the starting materials via conventional chemical–physical separations. Racemic acids can be resolved by reacting with an achiral alcohol or amine to form a chiral ester or amide, respectively.

The lipase from *C. cylindracea* has become a versatile catalyst for the resolution of racemic carboxylic acids and alcohols. For example, (±)-menthol can be resolved by the enzyme-catalyzed esterification of the (−)-isomer with lauric acid in heptane, and in optical purities in excess of 95% [172]. The (+)-isomer can be recovered as free menthol. Similar resolutions were performed with the same lipase entrapped in polyurethane gels, using 5-phenylpropionate in isooctane [173]. Gram-scale resolutions were afforded. Lipase from *C. cylindracea* has been effective in the preparative resolution of 2-halocarboxylic acids [123]. Gram quantities of 2-chloro- and 2-bromopropionic acids with optical purities in excess of 95% were prepared. Racemic aromatic acids have also been resolved in organic media [154]. The most important of these was 2-(4-chlorophenoxy)propionic acid, the *R*-isomer having use as a potent herbicide [174].

Porcine pancreatic lipase has proved effective in the optical resolution of racemic alcohols. For example, 2-octanol, 2-dodecanol, and *sec*-phenthanol were resolved by transesterification with trichloroethyl butyrate [123]. Only the *R*-isomer reacted with the enzyme, affording an *R*-based ester and optical resolution. The residual *S*-alcohol could be recovered easily by distillation. Acylation can be performed using lipases in the presence of an anhydride [175]. Inactivation due to functional group acylation was not evident. In another example of the power of enzyme-catalyzed optical resolution of alcohols, stereospecific esterification of the hydroxyl moiety of aminoalcohols is feasible by blocking the amine of 2-aminoalcohols [176]. The blocked amine is unable to act as the nucleophile, and esters form in the absence of amides.

In addition to aliphatic alcohols, optical resolution of organometallic alcohols and aliphatic amines has been carried out. In the former case, racemic ferrocenylethanol was resolved in toluene using porcine pancreatic lipase [142]. In the latter case, subtilisin in 3-methyl-3-pentanol was shown to resolve racemic amines with enantiomeric excess above 80% [177]. Solvent polarity was vital in determining the efficiency of optical separations. Nonpolar solvents such as toluene and cyclohexane did not provide a medium capable of sustaining enzymatic resolution of the amines. This is similar to the relaxation in enantioselectivity of various proteolytic enzymes afforded by hydrophobic solvents for the transesterification of amino acid derivatives (see above). Enzyme-catalyzed resolution or racemic mixtures is summarized in Table 5.

Miscellaneous Reactions

A number of reactions of interest do not fit the preceding categories. Proparglyamides have been prepared using porcine pancreatic and *C. cylindracea* lipases in carbon tetrachloride. Reaction of *p*-methoxyaniline with ethyl propynoate yielded the proparglyamide product. Conventional chemical synthesis yields the Michaels addition to the triple bond [182].

Carbohydrases have been used in monophasic organic solvents. Glucose condensation catalyzed by the reverse reaction of fungal glucoamylase in 90% diethylene glycol diethyl ether was performed [183]. A number of disaccharides (maltose and isomaltose) and trisaccharides (panose and isomaltotriose) were obtained from a 2% (w/v) solution of glucose. Invertase has been shown to reverse its sucrose hydrolysis activity in 70% organic solvent solutions including *tert*-butanol, acetone, and dioxane [184]. The synthesis of glycosides has been attempted in benzyl alcohol [185]. Low yields (generally < 10%) of benzyl glucoside were obtained using β-glucosidase covalently bound to a macroporous polyethylene matrix.

In addition to the poor solubility of sugars in apolar media, it has been suggested that polar substrates in organic media give tightly bound enzyme-sub-

Table 4 Representative Summary of Enzyme-Catalyzed Peptide Synthesis in Monophasic Organic Solvents[a]

Donor	Acceptor	Enzyme	Solvent	Product	Yield (5)	Ref.
Ac-Phe—OMe	Ala-NH$_2$	Chymotrpysin	50% DMF	Ac-Phe-Ala-NH$_2$	97	160
Ac-Phe—OMe	Ala-NH$_2$	Chymotrpysin	90% Butanediol	Ac-Phe-Ala-NH$_2$	88	160
Ac-Tyr	Gly-NH$_2$	Chymotrpysin	95% Acetonitrile	Ac-Tyr-Gly-NH$_2$	38	126
Z-Asp	Phe-OMe	Thermolysin	Ethyl acetate	Z-Asp-Phe-OMe	60	159
Z-Gly-Gly-Phe	Phe-NH$_2$	Thermolysin	tert-Amyl alcohol[c]	Z-Gly-Gly-Phe-Phe-NH$_2$	76	28
Z-Gly-Gly-Phe	Phe-Phe-NH$_2$	Thermolysin	tert-Amyl alcohol[c]	Z-Gly-Gly-Phe-Phe-Phe-NH$_2$	72	28
Z-Gly-Phe-Pro-Leu	Leu-NH$_2$	Thermolysin	tert-Amyl alcohol[c]	Z-Gly-Pro-Phe-Pro-Leu-Leu-NH$_2$	73	28
Z-Gly-Pro-Gly-Gly-Pro-Ala	Leu-Leu-Phe-NH$_2$	Thermolysin	tert-Amyl alcohol[d]	Z-Gly-Pro-Gly-Gly-Pro-Ala-Leu-Leu-Phe-NH$_2$	67	28
Z-Phe-OCH$_2$CN	Leu-NH$_2$	Chymotrpysin[e]	50% DMSO	Z-Phe-Leu-NH$_2$	88	167

Z-D-Phe-OCH₂CN	Leu-OMe	Chymotrpysin[e]	50% DMSO	Z-D-Phe-Leu-OMe	91	167
Z-Tyr-Gly-Gly-Phe-OCH₂CN	Leu-NH₂	Chymotrpysin[e]	50% DMSO	Z-Tyr-Gly-Gly-Phe-Leu1-NH₂	99	167
Pha-Cys(SAcm)-OMe	D-Val-OMe	Chymotrypsin	50% DMSO	Pha-Cys(SAcm)-D-Val-OMe	72	168
Z-Adp(OMe)-OH	Cys(SBzl)-D-Val-OBzl	Papain	50% DMSO	Z-Adp-Cys(SBzl)-D-Val-OBzl	42	168
Z-Tyr-OMe	D-Arg-OMe	Chymotrypsin	50% acetonitrile	Z-Tyr-D-Arg-OMe	72	168
Ac-Phe-OEtCl	Lys-Otert-Bu	Subtilisin	tert-Amyl alcohol[f]	Ac-Phe-Lys-Otert-Bu	85	169
F-D-Ala-OEtCl	Phe-NH₂	Subtilisin	DMF[f]	F-D-Phe-Phe-NH₂	82	70
F-D-Ala-OEtCl	Phe-Leu-NH₂	Subtilisin	DMF[f]	F-D-Ala-Phe-Leu-NH₂	61	70

[a]Unless otherwise noted, L-amino acids are employed.
[b]Abbreviations: Z, benzyloxycarbonyl; Pha, phenylacetyl; Acm, acetamidomethyl; Bzl, benzyl; Adp, α-aminoadipic acid.
[c]Containing 1% water and 9% formamide.
[d]Containing 1% water and 9% ethylene glycol.
[e]Chymotrypsin chemically methylated at $N^{\epsilon2}$ of His57.
[f]Anhydrous solvent.

Table 5 Representative Summary of Optical Resolutions Employing Enzymes in Monophasic Organic Solvents

Racemate	Acyl donor or acceptor	Enzyme	Solvent	Isomer reacted	Yield[a]	EE[b]	Ref.
Menthol	Lauric acid	Lipase from *Candida cylindracea*	Heptane	(−)	45	95	172
2-Decanol	*n*-Butyl alcohol	Lipase from *Mucor miehei*	Hexane	R	40	87	178
2-Chloropropionic acid	*n*-Butyl alcohol	Lipase from *C. cylindracea*	Hexane	R	42	95	123
2-Bromopropionic acid	*n*-Butyl alcohol	Lipase from *C. cylindracea*	Hexane	R	45	96	123
2-Octanol	Trichloroethyl-butyrate	Lipase from porcine pancreas	Diethyl ether	R	47	95	123
2-Methyl-2,4-pentanediol	Trichloroethyl-butyrate	Lipase from porcine pancreas	Heptane	R	48	87	123
Butyl-O-formyl-mandelate	*n*-Butyl alcohol	Lipase from *C. cylindracea*	Isopropyl ether	R	39	75	179
2-Octanol	Dodecanoic acid	Lipase from porcine pancreas	Heptane	R	43	97	180
2-(4-Chloro)-propionic acid	Cyclohexanol	Lipase from *C. cylindracea*	Isoctane	R	40	98	154

Menthol	Lauric acid	Lipase from C. cylindracea	Isoctane	R	40	99	154
2-N-Alk-amino-propanol[c]	Ethyl acetate	Lipase from porcine pancreas	Ethyl acetate	R	32[d]	92	154
2-N-Alk-amino-propanol	Ethyl acetate	Pancreatin	Ethyl acetate	R	25[d]	95	176
2-N-Alk-amino-propanol	Ethyl acetate	Lipase from Pseudomonas sp.	Ethyl acetate	R	14[d]	85	176
1,1-Binaphthyl-2,2'-diol	Vinyl acetate	Lipase from Pseudomonas sp.	Isopropyl ether	R	53	95	181
2-O-Benzylglycerol	Isopropenyl acetate	Lipase from Pseudomonas sp.	Chloroform	S	36	96	142
Ferrocenylethanol	Vinyl propionate	Lipase from Pseudomonas sp.	Toluene	S	30	84	142
α-methyl benzyl-amine	Trifluorethyl-butyrate	Subtilisin	3-Methyl-3-pentanol	S	35	85	177

[a]Unless otherwise stated, this represents a percentage of the reacting isomer.
[b]Enantiomeric excess.
[c]Alk, alkoxycarbonyl.
[d]Isolated yield of recovery of 2-aminopropanol following alkaline hydrolysis.

strate or enzyme–product complexes [186]. Low dissociation rates of these enzyme complexes lead to minimal catalytic rates and conversions. Addition of water mimics such as formamide may alleviate complex formation, but this has yet to be proved.

The first demonstration of enzymatic activity in supercritical fluids was achieved using alkaline phosphatase in CO_2 for the hydrolysis of disodium p-nitrophenyl phosphate to p-nitrophenol [187]. The enzyme retained full activity after 24 hours in CO_2 at 100 bar and 35°C. Because of the poor solubility of the substrate, however, the observed catalysis probably occurred in a thin water film surrounding the enzyme.

Isomerases

The only known application of isomerases in monophasic organic solvents is the use of glucose (xylose) isomerase in ethanol–water mixtures [188]. In aqueous solutions, glucose isomerization typically leads to a mixture of 42% fructose, 51% glucose, and 7% oligosaccharides (that remain from the incomplete amylase-catalyzed hydrolysis of starch) [189]. Commercial high fructose corn syrup (HFCS) requires 55% fructose and is achieved through ion-exchange enrichment [190]. One potential solution for fructose enrichment without chromatography involves the use of glucose isomerase in 80% ethanol. In this solvent, the enzyme retains 90% of its aqueous-based activity, and the equilibrium concentration of fructose is increased to 55%.

FUTURE DIRECTIONS

Enzymatic catalysis in nonaqueous media is already having an impact in the way biochemists, chemists, and engineers view applied enzymology. Biotechnologists are no longer limited to aqueous solutions for biocatalysis. Chemists may now take advantage of enzymic specificity under mild operating conditions to catalyze reactions formerly limited to expensive and tedious chemical processes. New reactions and processes using enzymes in organic solvents are being developed at an accelerated pace.

The full application and acceptance of this novel technology, however, requires the continuation of several research directions. One top priority is to examine quantitatively the nature of organic solvents on enzyme structure and function. Mathematical relationships must be developed that can successfully predict the effect of a given organic solvent on an enzyme's catalytic activity and substrate specificity. Such a research program will invariably involve protein engineering to "design" proteins with specific physicochemical properties deemed optimal for organic solvent function. Another important direction that must be examined is the biochemical engineering of enzymes in organic solvents. Reactors designed specifically for nonaqueous enzymology must take into account the precise control of water activity, biocatalyst preparation on insoluble supports, and the compatibility of the organic solvent based enzyme process with upstream, downstream, and potentially linked chemical processing. Finally, as reactions conventionally carried out with chemical catalysts give way to highly selective biocatalysts, new research directions will be uncovered. The field of nonaqueous enzymology has matured into a discipline of biocatalysis that will significantly expand the use of biotechnology in the chemical, polymer, and pharmaceutical industries.

ACKNOWLEDGMENTS

The author wishes to acknowledge Keungarp Ryu and Dr. Zu-Feng Xu for their help in searching the literature.

REFERENCES

1. Borgstrom, B., and Brockman, H. L. (eds.). *Lipases*, Elsevier, Amsterdam, 1984.
2. Gunsales, I. C., Meeks, J., Lipscomb, J., Debrunner, P., and Munck, E. In *Molecular Mechanisms of Oxygen Activation* (O. Hayaishi, ed.), Academic Press, New York, pp. 561–614, 1974.
3. Dordick, J. S. *Enzyme Microb. Technol. 11*, 194–211 (1989).
4. Klibanov, A. M. *Chemtech, 16*, 354–359 (1986).
5. Butler, L. G. *Enzyme Microb. Technol. 1*, 253–259 (1979).
6. Lilly, M. D. *J. Chem. Tech. Biotechnol. 32*, 162–169 (1982).
7. Carrea, G. *Trends Biotechnol. 2*, 102–106 (1984).
8. Halling, P. J. *Biotechnol. Adv. 5*, 47–84 (1987).
9. Martinek, K., Levashov, A. V., Kylachko, N., Khmelnitski, Y. L., and Berezin, I. V. *Eur. J. Biochem. 155*, 453–468 (1986).
10. Luisi, P. L. and Laane, C. *Trends Biotechnol. 4*, 153–161 (1986).
11. Tanford, C. *Physical Chemistry of Macromolecules*, Wiley, New York, 1961.
12. Schulz, G. E., and Schirmer, R. H. *Principles of Protein Structure*, Springer-Verlag, New York, 1979.
13. Careri, G., Gratton, E., Yang, P.-H., and Rupley, J. A. *Nature (London) 284*, 572–573 (1980).
14. Rupley, J. A., Gratton, E., and Careri, G. *Trends Biochem. Sci. 8*, 18–22 (1983).
15. Schinkel, J. E., Downer, N. W., and Rupley, J. A. *Biochemistry, 24*, 352–366 (1985).
16. Zaks, A., and Klibanov, A. M. *J. Am. Chem. Soc. 108*, 2767–2768 (1986).
17. Therisod, M., and Klibanov, A. M. *J. Am. Chem. Soc. 108*, 5638–5640 (1986).
18. Therisod, M., and Klibanov, A. M. *J. Am. Chem. Soc. 109*, 3977–3981 (1987).

19. Zaks, A., and Klibanov, A. M. *J. Biol. Chem. 263*, 3194-3201 (1988).
20. Kazandjian, R. Z., and Klibanov, A. M. *J. Am. Chem. Soc. 107*, 5448-5450 (1985).
21. Kazandjian, R. A., Dordick, J. S., and Klibanov, A. M. *Biotechnol. Bioeng. 28*, 417-421 (1986).
22. Zaks, A., and Klibanov, A. M. *J. Biol. Chem. 263*, 8017-8021 (1988).
23. Almagor, E., and Belfort, G. *J. Colloid Interface Sci. 66*, 146-152 (1978).
24. Angell, C. A. In *Water, A Comprehensive Treatise*, Vol. 7 (F. Franks, ed.), Plenum Press, New York, 1982, pp. 1-81.
25. Wiggins, P. M., and van Ryn, R. T. *J. Macromol. Sci. Chem. A23*, 975-983 (1986).
26. Wiggins, P. M. *Prog. Polym. Sci. 13*, 1-35 (1988).
27. O'Connor, C. J., and Wiggins, P. M. *Biocatalysis, 1*, 249-256 (1988).
28. Kitaguchi, H., and Klibanov, A. M. *J. Am. Chem. Soc. 111*, 9272-9273 (1989).
29. Laane, C., Tramper, J., and Lilly, M. D. (eds.). *Biocatalysis in Organic Media*, Elsevier, Amsterdam, 1987.
30. Wasacz, F. M., Olinger, J. M., and Jakobsen, R. J. *Biochemistry, 26*, 1464-1470 (1987).
31. Dordick, J. S. Ph.D. dissertation, Massachusetts Institute of Technology, 1986.
32. Herskovits, T. T. *J. Biol. Chem. 240*, 628-644 (1965).
33. Tanford, C. *J. Am. Chem. Soc. 84*, 4240-4247 (1962).
34. Klotz, I. M., and Franzen, J. S. *J. Am. Chem. Soc. 82*, 5241-5242 (1960).
35. Weber, R. E., and Tanford, C. *J. Am. Chem. Soc. 81*, 3255 (1959).
36. Klysov, A. A., Van Niet, N., and Berezin, I. V. *Eur. J. Biochem. 59*, 3-7 (1975).
37. Herskovits, T. T., Gadegbeku, B., and Jaillet, H. *J. Biol. Chem. 245*, 2588-2598 (1970).
38. Ryu, K., Stafford, D. R., and Dordick, J. S. *ACS Symp. Ser. 389*, 141-157 (1989).
39. Clark, D. S., Skerker, P. S., Creagh, L., Guinn, M., Prausnitz, J., and Blanch, H. *ACS Symp. Ser. 392* (1989).
40. Skerker, P. S., and Clark, D. S. *Biotechnol. Bioeng. 33*, 62-71 (1989).
41. Zaks, A., and Klibanov, A. M. *Proc. Natl. Acad. Sci. USA, 82*, 3192-3196 (1985).
41a. Patil, D. R.., Rethwisch, D. G., and Dordick, J. S., *Biotechnol. Bioeng.* (in press).
42. Zaks, A., and Klibanov, A. M. *Science, 224*, 1249-1251 (1984).
43. Russell, A. J., and Klibanov, A. M. *J. Biol. Chem. 263*, 11624-11626 (1988).
44. Braco, L., Dabulis, K., and Klibanov, A. M. *Proc. Natl. Acad. Sci. USA, 87*, (1990).
45. Russell, A. J., Trudel, L. J., Skipper, P. L., Groopman, J. D., Tannenbaum, S. R., and Klibanov, A. M. *Biochem. Biophys. Res. Commun. 158*, 80-85 (1989).
46. Tanford, C. *Adv. Protein Chem. 23*, 121-282 (1968).
47. Feeney, R. E. In *Chemical Deterioration of Proteins* (J. R. Whitaker and M. Fujimaki, eds.), American Chemical Society, Washington, DC, 1980, p. 1.
48. Ayala, G., Tuena de Gomez-Puyou, M., Gomez-Puyou, A., and Darszon, A. *FEBS Lett. 203*, 41-43 (1986).
49. Ueda, M., Mukatka, S., Sato, S., and Takahashi, T. *Agric. Biol. Chem. 50*, 1533-1537 (1986).

50. Jenks, W. P. *Adv. Enzymol. 43*, 219 (1975).

51. Kraut, J. *Science, 242*, 533-540 (1988).

52. Fersht, A. *Enzyme Structure and Mechanism*, 2nd ed., Freeman, New York, 1985.

53. Kaufman, S., and Neurath, H. *J. Biol. Chem. 180*, 181-187 (1949).

54. Barnard, M. L., and Laidler, K. J. *J. Am. Chem. Soc. 74*, 6099-6101 (1952).

55. Miles, J. L., Moley, E., Crain, F., Gross, S., San Julian, J., and Canady, W. J. *J. Biol. Chem. 237*, 1319-1322 (1962).

56. Bender, M. L. *Methods Enzymol. 135*, 537-546 (1987).

57. Bender, M. L., and Kezdy, F. J. *J. Am. Chem. Soc. 86*, 3704 (1964).

58. Manrel, P., *J. Bio. Chem. 253* 1677-1683 (1978).

59. Ryu, K., and Dordick, J. S. *J. Am. Chem. Soc. 111*, 8026-8027 (1989).

60. Saunders, B. C., Holmes-Siedle, A. G., and Stark, B. P. *Peroxidase*, Butterworths, Washington, DC, 1964.

61. Laane, C., Boeren, S., Vos, K., and Veeger, C. *Biotechnol. Bioeng. 30*, 81-87 (1987).

62. Douzou, P., and Balny, C. *Proc. Natl. Acad. Sci. USA, 74*, 2297-2300 (1977).

63. Wells, P. R. *Linear Free Energy Relationships*, Academic Press, New York, 1968.

64. Bender, M. L., and Nakamura, K. J. *J. Am. Chem. Soc. 84*, 2577-2582 (1962).

65. Hubbard, C. D., and Shoupe, T. S. *J. Biol. Chem. 252*, 1633-1638 (1977).

66. Ikeda, K., Kunugi, S., and Ise, N. *Arch. Biochem. Biophys. 217*, 37-46 (1982).

67. Kanerva, L. T., and Klibanov, A. M. *J. Am. Chem. Soc. 11*, 6864-6865 (1989).

68. Burke, P. A., Smith, S. O., Bachovchin, W. W., and Klibanov, A. M. *J. Am. Chem. Soc. 111*, 8290-8291 (1989).

69. Jones, J. B., and Mehes, M. M. *Can. J. Chem. 57*, 2245-2248 (1979).

70. Margolin, A. L., Tai, D.-F., and Klibanov, A. A. *J. Am. Chem. Soc. 109*, 7885-7887 (1987).

71. Sakurai, T., Margolin, A. L., Russell, A. J., and Klibanov, A. M. *J. Am. Chem. Soc. 110*, 7236-7237 (1988).

72. Laane, C., Boeren, S., Hilhorst, R., and Veeger, C. In *Biocatalysis in Organic Media* (C. Laane, J. Tramper, and M. D. Lilly, eds.), Elsevier, Amsterdam, 1987, pp. 65-84.

73. Cantor, C. R., and Schimmel, P. R. *Biophysical Chemistry*, Vol. 2, Freeman, New York, 1980.

74. Takahashi, K., Nishimura, H., Yoshimoto, T., Saito, Y., and Inada, Y. *Biochem. Biophys. Res. Commun. 121*, 261-265 (1984).

75. Takahashi, K., Ajima, A., Yoshimoto, T., and Inada, Y. *Biochem. Biophys. Res. Commun. 125*, 761-766 (1984).

76. Takahashi, K., Ajima, A., Yoshimoto, T., Okada, M., Matsushima, A., Tamaura, Y., and Inada, Y. *J. Org. Chem. 50*, 3414-3415 (1985).

77. Martinek, K., Levashov, A. V., Khmelnitski, Y. L., Klysov, N. L., and Berezin, I. V. *Science, 218*, 889-891 (1982).

78. Dordick, J. S., Marletta, M. A., and Klibanov, A. M. *Proc. Natl. Acad. Sci. USA, 83*, 6255-6257.

79. Tramper, J. *Trends Biotechnol. 3*, 45-50 (1985).
80. Reslow, M., Aldercreutz, P., and Mattiasson, B. *FEBS*, 573-578 (1988).
81. Yokozeki, K., Tamanaka, T., Utagawa, T., Takinami, K., Hirose, Y., Tanaka, A., Sonomoto, K., and Fukui, S. *Eur. J. Appl. Microbiol. Biotechnol. 14*, 225-231 (1982).
82. Chiltenden, R. H., and Mendel, L. B. *Am. J. Med. Sci. 3*, 181-186 (1896).
83. Bayliss, W. M. *J. Physiol. 50*, 85-94 (1915).
84. Dastoli, F. R., and Price, S. *Arch. Biochem. Biophysd. 118*, 163-165 (1967).
85. Siegel, S. M., and Roberts, K. *Space Life Sci. 1*, 131-134 (1968).
86. Sedlaczek, L. *CRC Crit. Rev. Biotechnol. 7*, 187-236 (1988).
87. Buckland, B. C., Dunnill, P., and Lilly, M. D. *Biotechnol. Bioeng. 17*, 815-826 (1975).
88. Randolph, T. W., Clark, D. S., Blanch, H. W., and Prausnitz, J. M. *Science, 238*, 387-390 (1988).
89. Randolph, T. W., Clark, D. S., Blanch, H. W., and Prausnitz, J. M. *Proc. Natl. Acad. Sci. USA, 85*, 2979-2983 (1988).
90. Omata, T. Iida, T., Tanaka, A., and Fukui, S. *Eur. J. Appl. Microbiol. Biotechnol. 8*, 143-155 (1979).
91. Fukui, S., and Tanaka, A. *Endeavor, 9*, 10-17 (1985).
92. Fukui, S. Ahmed, S. A., Omata, T., and Tanaka, A. *Eur. J. Appl. Microbiol. Biotechnol. 10*, 289-301 (1980).
93. Wong, C.-H., and Whitesides, G. M. *J. Am. Chem. Soc. 103*, 4890-4899 (1981).
94. Grunwald, J., Wirtz, B., Scollar, M. P., and Klibanov, A. M. *J. Am. Chem. Soc. 108*, 6732-6734 (1986).
95. Deetz, J. S., and Rozzell, J. D. *Trends Biotechnol. 6*, 15-19 (1988).
96. POulvin, S., Legoy, M. D., Lorde, R., Pensa, M., and Thomas, D. *Biotechnol. Lett. 8*, 783-784 (1986).
97. Sarkanen, K. V., and Ludwig, C. H. (eds.). *Lignins: Occurrence, Formation Structure, and Reactions*, Parts 1 and 2, Wiley, New York, 1971.
98. Brode, G. L. In *Kirk-Othmer Encyclopedia of Chemical Technology*, 3rd ed., Vol. 17, Wiley, New York, 1981, pp. 384-416.
99. Marshall, E. *Science, 237*, 381 (1987).
100. Schwartz, R. D., and Hutchinson, D. B. *Enzyme Microb. Technol. 3*, 361-363 (1981).
101. Dordick, J. S., Marletts, M. A., and Klibanov, A. M. *Biotechnol. Bioeng. 30*, 31-36 (1987).
102. Whitehouse, A. A. K., Pritchett, E. G. K., and Barnett, G. *Phenolic Resins*, Elsevier, New York, 1968, Chapter 2.
103. Pokora, A. R., and Cyrus, W. L. U.S. Patent 4,647,952 (1987).
104. Boeriu, C. G., Dordick, J. S., and Klibanov, A. M. *Biotechnology, 4*, 997-999 (1986).
105. Klibanov, A. M., and Dordick, J. S. U.S. Patent 4,826,762 (1989).
106. Brink, L. E. S., and Tramper, J. *Enzyme Microb. Technol. 8*, 281-288 (1986).
107. Takazawa, Y., Sato, S., and Takahashi, J. *Agric. Biol. Chem. 48*, 2489-2495 (1984).

108. Hammond, D. A. Karel, M., Klibanov, A. M., and Krukonis, V. J. *Appl. Biochem. Biotechnol. 11*, 393-400 (1985).
109. Hyatt, J. A. *J. Org. Chem. 49*, 5097-5101 (1984).
110. Milstein, O., Nicklas, B., Trojanowski, J., Majcherczyk, A., and Huttermann, A. *TAPPI Proc.* 317-324 (1989).
111. Milstein, O., Nicklas, B., and Huttermann, A. *Appl. Microbiol. Biotechnol. 31*, 70-74 (1989).
112. Brumm, P. J. *Biotechnol. Lett. 10*, 237-242 (1988).
113. Barzana, E., Klibanov, A. M., and Karel, M. *Appl. Biochem. Biotechnol. 15*, 25-34 (1986).
114. Barzana, E. Klibanov, A. M., and Karel, M. *Biotechnol. Bioeng. 34*, 1178-1185 (1989).
115. Chambert, R., and Petit-Glatron, M. F. *Carbohydr. Res. 191*, 117-123 (1989).
116. Bell, G., Blain, J. A., Paterson, J. D. E., Shaw, C. E. L., and Todd, R. J. *FEMS Microbiol. Lett. 3*, 223-225 (1978).
117. Paterson, J. D. E., Blain, J. A., Shaw, C. E. L., and Todd, R. J. *Biotechnol. Lett. 1*, 211-216 (1979).
118. Marlot, C., Langrand, G., Triantaphylides, C., and Baratti, J. *Biotechnol. Lett. 7*, 647-650 (1985).
119. Koshiro, S., Sonomoto, K., Tanaka, A., and Fukui, S., *J. Biotechnol. 2*, 47-57 (1985).
120. Gatfield, I., and Sand, T. U.S. Patent 4,451,565 (1984).
121. Okumura, S., Iwai, M., and Tsujisaka, Y. *Biochem. Biophys. Acta, 575*, 156-165 (1987).
122. Hiratake, J., Inagaki, M., Nishioka, T., and Oda, J. *J. Org. Chem. 53*, 6130-6133 (1988).
123. Kirchner, G., Scollar, M. P., and Klibanov, A. M. *J. Am. Chem. Soc. 107*, 7072-7076 (1985).
124. van Eijs, A. M. M., Oostrom, W. H. M., and Wijsman, J. A. *Proceedings of the 4th European Congress on Biotechnology*, Vol. 2, 1987, p. 211.
125. Kise, H., and Shirato, H. *Tetrahedron Lett. 15*, 6081-6084 (1985).
126. Kise, H., Hayakawa, A., and Noritomi, H. *Biotechnol. Lett. 9*, 543-548 (1987).
127. Turkova, J., Vanek, T., Turkova, R., Veruovic, B., and Kubanek, V. *Biotechnol. Lett. 3*, 165-170 (1982).
128. Tai, D.-F., Fu, S.-L., Chuang, S.-F., and Tsai, H. *Biotechnol. Lett. 11*, 173-176 (1989).
129. Chen, C.-T., and Wang, K. T. *J. Chem. Soc. Chem. Commun.* 327-328 (1988).
130. Weetall, H. H. *Biotechnol. Bioeng. 27*, 124-127 (1985).
131. Pasta, P., Mazzola, G., Carrea, G., and Riva, S. *Biotechnol. Lett. 11*, 643-648 (1989).
132. Notheisz, F., Bartok, M., and Remport, V. *Acta Phys. Chem. 18*, 89-98 (1972).
133. Mosher, H. S., and Morrison, J. D. *Science, 221*, 1013-1019 (1983).
134. Haines, A. H. *Adv. Carbohydr. Chem. Biochem. 39*, 13-70 (1981).
135. Desgupta, F., Hay, G. W., Szarek, W. A., and Schilling, W. L. *Carbohydr. Res. 114*, 153-157 (1983).

136. Cesti, P., Zaks, A., and Klibanov, A. M. *Appl. Biochem. Biotechnol. 11*, 401-407 (1985).
137. Chinsky, N., Margolin, A. L., and Klibanov, A. M. *J. Am. Chem. Soc. 111*, 386-388 (1989).
138. Watanabe, K. A., and Fox, J. J. *Angew. Chem. 78*, 589 (1966).
139. Riva, S., and Klibanov, A. M. *J. Am. Chem. Soc. 110*, 3291-3295 (1988).
140. Bjorkling, F., Godtfredson, S. E., and Kirk, O. *J. Chem. Soc. Chem. Commun.* 934-935 (1989).
141. Chopineau, J., McCafferty, F. D., Therisod, M., and Klibanov, A. M. *Biotechnol. Bioeng. 31*, 208-214 (1988).
142. Wang, Y.-F., Lalonde, J. J., Momongan, M., Bergbreiter, D. E., and Wong, C.-H. *J. Am. Chem. Soc. 110*, 7200-7205 (1988).
143. Dordick, J. S., Hacking, A. J., and Khan, R. A. U.S. Patent Pending (1990).
144. Carrea, G., Riva, S., Secundo, F., and Danieli, B. *J. Chem. Soc. Perkin Trans.* 1057-1061 (1989).
145. Riva, S., Chopineau, J., Kieboom, A. P. G., and Klibanov, A. M. *J. Am. Chem. Soc. 110*, 584-589 (1988).
146. Matsuo, T. Sawamura, N., Hashimoto, Y., and Hashida, W. U.S. Patent 4,268,527 (1981).
147. Eigtved, P. Hansen, T. T., and Huge-Jensen, B. *Abstracts of the 13th Scandinavian Symposium on Lipids*, Reykjavik, 1985.
148. Macrae, A. R. *J. Assoc. Off. Anal. Chem. 60*, 243-246 (1983).
149. Goderis, H. L. Feyten, M. R., Fouwe, B. L., Guffens, W. M., van Cauwenbergh, S. M., and Toback, P. P. *Biotechnol. Bioeng. 30*, 258-266 (1987).
150. Gatfield, I. *Ann. N.Y. Acad. Sci. 434*, 569 (1984).
151. Gutman, A. L., Zuobi, K., and Boltansky, A. *Tetrahedron Lett. 28*, 386-387 (1987).
152. Makita, A., Nihira, T., and Yamada, Y. *Tetrahedron Lett. 28*, 805-808 (1987).
153. Zhi-Wei, G., and Sih, C. J. *J. Am. Chem. Soc. 110*, 1999-2001 (1988).
154. Chen, C.-S., Wu, S.-H., Girdaukas, G., and Sih, C. J. *J. Am. Chem. Soc. 109*, 2812-2817 (1987).
155. Ajima, A., Yoshimoto, T., Takahashi, K., Tamaura, Y., Saito, Y, and Inada, Y. *Biotehcnol. Lett. 7*, 303-306 (1987).
156. Margolin, A. L., Crenne, J.-Y., and Klibanov, A. M. *Tetrahedron Lett. 28*, 1607-1610 (1987).
157. Wallace, J. S., and Morrow, C. J. *J. Polym. Sci. Part A: Polym. Chem. 27*, 2553-2567 (1989).
158. Abramowicz, D. A., and Keese, C. R. *Biotechnol. Bioeng. 33*, 149-156 (1989).
159. Ooshima, H., Mori, H., and Harano, Y. *Biotechnol. Lett. 7*, 789-792 (1985).
160. Nilsson, K., and Mosbach, K. *Biotechnol. Bioeng. 26*, 1146-1154 (1984).
161. Jakubke, H. D., Kuhl, P., and Konnecke, A. *Angew. Chem. Int. Ed. Engl. 24*, 85-93 (1985).
162. Cassells, J. M., and Halling, P. J. *Enzyme Microb. Technol. 10*, 486-491 (1988).
163. Ferjancic, A., Puigserver, A., and Gaertner, H. *Biotechnol. Lett. 10*, 101-106 (1988).

164. Oyama, K. In *Biocatalysis in Organic Media* (C. Laane, J. Tramper, and M. D. Lilly, eds.), Elsevier, Amsterdam, 1987, pp. 209-224.
165. Margolin, A. L., and Klibanov, A. M. *J. Am. Chem. Soc. 109*, 3802-3804 (1987).
166. West, J. B., and Wong, C.-H. *Tetrahedron Lett. 28*, 1629-1632 (1987).
167. West, J. B., Scholten, J., Stolowich, N. J., Hogg, J. L., Scott, A. I., and Wong, C.-H. *J. Am. Chem. Soc. 110*, 3709-3710 (1988).
168. Barbas, C. F., Matos, J. R., West, J. B., and Wong, C.-H. *J. Am. Chem. Soc. 110*, 5162-5166 (1988).
169. Kitaguchi, H., Tai, D.-F., and Klibanov, A. M. *Tetrahedron Lett. 29*, 5487-5488 (1988).
170. Marshek, W. J., and Miyano, M. *Biochim. Biophys. Acta, 316* 363-365 (1973).
171. Ladner, W. E., and Whitesides, G. M. *J. Am. Chem. Soc. 106*, 7250-7251 (1984).
172. Langrand, G., Baratti, J., Buono, G., and Triantaphylides, C. *Tetrahedron Lett. 27*, 29-32 (1986).
173. Yokozeki, K., Yamanaka, S., Takinami, K., Hirose, Y., Tanaka, A., Sonomoto, K., and Fukui, S. *Eur. J. Appl. Microbiol. Biotechnol. 14*, 1-5 (1982).
174. Cremlyn, R. *Pesticides, Preparation and Mode of Action*, Wiley, Chichester, 1978, p. 143.
175. Bianchi, D., Cesti, P., and Battistel, E. *J. Org. Chem. 53*, 5531-5534 (1988).
176. Francalanci, F., Cesti, P., Cabri, W., Bianchi, D., Martinengo, T., and Foa, M. *J. Org. Chem. 52*, 5079-5082 (1987).
177. Kitaguchi, H., Fitzpatrick, P. A., Huber, J. E., and Klibanov, A. M. *J. Am. Chem. Soc. 111*, 3094-3095 (1989).
178. Sonnet, P. E. *J. Org. Chem. 52*, 3477-3479 (1987).
179. Bevinakatti, H. S., and Newadkar, R. V. *Biotechnol. Lett. 11*, 785-788 (1989).
180. Gerlach, D., and Schreier, P. *Biocatalysis, 2*, 257-263 (1989).
181. Inagaki, M., Hiratake, J., Nishioka, T., and Oda, J. *Agric. Biol. Chem. 53*, 1879-1884 (1989).
182. Rebolledo, F., Brieva, R., and Gotor, V. *Tetrahedron Lett. 30*, 5345-5346 (1989).
183. Laroute, V., and Willemot, R.-M. *Biotechnol. Lett. 11*, 249-254 (1989).
184. Straathof, A. J. J., Vrijenhoef, J. P., Sprangers, E. P. A. T., van Bekkem, H., and Kieboom, A. P. G. *J. Carbohydr. Chem. 7*, 223-238 (1988).
185. Marek, M., Novotna, Z., Jary, J., and Kocikova, V. *Biocatalysis, 2*, 239-243 (1989).
186. Kieboom, A. P. G. *Recl. Trav. Chim. Pays-Bas, 107*, 347-348 (1988).
187. Randolph, T. W., Blanch, H. W., Prausnitz, J. M., and Wilke, C. R. *Biotechnol. Lett. 7*, 325-328 (1985).
188. Visuri, K., and Klibanov, A. M. *Biotechnol. Bioeng. 30*, 917-920 (1987).
189. Bucke, C. In *Topics in Enzyme and Fermentation Biotechnology*, Vol. 1 (A. Wiseman, ed.), Ellis Horwood, Chichester, 1977, pp. 147-171.
190. Antrim, R. L., Colilla, W., and Schnyder, B. P. In *Applied Biochemistry and Bioengineering*, Vol. 2 (L. Wingard, E. Katchalski-Katzir, and L. Goldstein, eds.), Academic Press, New York, 1979, pp. 97-155.

2

Enzyme Design for Nonaqueous Solvents

Jack Y. Hwang and Frances H. Arnold

California Institute of Technology, Pasadena, California

INTRODUCTION

The ability to use enzymes in nonaqueous solvents greatly expands the potential scope and economic impact of biocatalysis. Enzymatic catalysis in nonaqueous solvents offers possibilities for new chemistry using reactions not feasible in aqueous media due to kinetic or thermodynamic constraints. Yields for reactions that are favored only at low concentrations of water (e.g., the synthesis of amide bonds by proteases) can be greatly enhanced by carrying out the synthesis in nonaqueous media in the absence of significant quantities of water. Practical applications of enzymes in water are severely limited by the poor solubility in that medium of most organic substrates. The large reactor volumes required and the deleterious effects on reaction kinetics reduce the effectiveness of dilute systems. Furthermore, since water participates in many of the reactions that lead to irreversible loss of enzyme activity [1], a nonaqueous medium may greatly enhance enzyme stability [2]. These and other potential advantages of carrying out enzymatic conversions in nonaqueous solvents have been reviewed elsewhere [2–5] and are discussed in Chapter 1 of this volume.

This chapter proposes a set of rules for enhancing the stability and activity of enzymes in nonaqueous solvents by protein engineering. It is argued that significant improvements in enzyme stability in organic solvents can be achieved through alterations in amino acid sequence. Based on consideration of the forces that contribute to the stability of folded proteins, we have attempted to predict the types of substitutions of amino acids in the protein sequence that are likely to improve enzyme stability and activity in nonaqueous solvents.

Numerous examples of enzymes that remain catalytically active in nonpolar organic solvents have appeared in the literature and are reviewed by Dordick [3]. Most enzymes will retain some activity when suspended in nonpolar solvents containing very little water. Highly nonpolar solvents are nonsolvents for proteins and do not interact with the enzyme or with water at the enzyme's surface. Enzymes in nonpolar solvents have in fact been compared to dry or partially hydrated enzymes [6,7]. However, nonpolar solvents are often poor solvents for substrates of interest. For example, the low solubility of the products of polymerization reactions drastically limits the polymer molecular weight that can be achieved before the product precipitates and is unavailable for further reaction. Many hydrophilic substrates, such as amino acids, are nearly insoluble in nonpolar media.

Polar organic solvents would provide a better reaction medium for many industrial and synthetic applications. Unfortunately, there is a correlation between the loss of enzyme activity in nonaqueous solvents and solvent polarity [8,9]. Polar solvents of practical interest, such as acetone or dimethylformamide, interact with the enzyme and associated water molecules and drastically reduce or obliterate catalysis. These solvents are strong protein denaturants and quickly lead to deactivation.

This negative response to polar solvents is not surprising. Selective evolutionary pressures were never applied to organisms to cause them to develop enzymes for use in polar organic solvents. An enzyme in a polar organic solvent is a system that has been severely perturbed from its optimum. The noncovalent interactions that determine the stability of a folded protein will attain a new balance in an organic solvent, and that balance can favor loss of catalytic activity and, in the case that it is kinetically accessible, unfolding. However, in such an unoptimized system there is considerable room for improvement. It should be possible to make modifications in the amino acid sequence of enzymes to redistribute noncovalent forces in order to compensate for the unfavorable changes brought about by the change in solvent. By compensating for some of the unfavorable effects through reengineering the protein sequence, one may be able to enhance enzyme activity and stability in polar organic solvents.

The situation is in some ways similar to engineering improved thermostability into enzymes. Specific mechanisms of enzyme thermostabilization have been proposed (see, e.g., refs. 10 and 11), and there have been several successful attempts at engineering improvements in thermostability by site-directed mutagenesis [12–16]. Because the free energy of protein folding is small (5–20 kcal/mole), and corresponds to only a few noncovalent interactions, minimal alterations in amino acid sequence can have dramatic effects on protein thermostability.

The design rules for engineering nonaqueous solvent-stable enzymes proposed in this chapter are based on studies of protein–solvent interactions and solvent effects on the noncovalent forces that determine protein folding and stability. The proposed rules are as yet largely untested and thus must be validated and reevaluated through future protein engineering studies. Before presenting the design principles, it is useful to review selected studies of solvent effects on protein structure and dynamics. As a point of departure, we will discuss studies of the sequential hydration of proteins from the dry state. These studies have attempted to determine how the hydration process affects protein conformation and flexibility as well as to elucidate the role of solvent in the onset of enzymatic activity. Although we are far from a complete understanding of the solvation processes that ultimately lead to enzyme function, these hydration studies may yield important clues to enhancing the activity of enzymes suspended in organic solvents.

The discussion of hydration is followed by a limited review of the interactions that contribute to protein stability in solution. This discussion of stabilizing interactions is accompanied by examples from the protein engineering literature of amino acid substitutions that either enhance or diminish protein stability. By considering how a nonaqueous solvent affects these interactions, we have attempted to identify those mechanisms that will be particularly important in retaining the stability of enzymes dissolved in nonaqueous media.

The final section synthesizes results from hydration and solution studies to formulate a set of design criteria for engineering nonaqueous solvent-stable enzymes. The connections between structural features and therefore the types of amino acid alterations that should result in improved stability in aqueous solution and stability in nonaqueous solvents will be explored. These design criteria are illustrated with a comparison of the sequences and structures of crambin, a protein that is stable in a wide variety of polar organic solvents, and its water-soluble homolog. Finally, amino acid substitutions that improve the stability of the protease subtilisin BPN' in organic solvents are discussed.

PROTEIN HYDRATION STUDIES

The role of water in activating a dry enzyme has been the subject of much research in recent years [17–21]. Spectroscopic techniques, which include difference IR[22], Raman measurements [23,24], NMR [25], and ESR [26], as well as thermodynamic measurements, which include heat capacities and enthalpies of hydration [19,26,27], have been used to probe the sequential hydration of proteins of known structure: for example, hen egg white lysozyme [18,28], bovine α-lactalbumin [19], myoglobin [20], and ribonuclease A [21]. Although there is no universal agreement in the interpretation of the data, a useful working model emerges as a result of these studies that provides insight into the role of water in effecting catalysis at a molecular level. We describe below some representative experiments on the hydration of lysozyme, laying special emphasis on those aspects of hydration that appear relevant to enzyme activity.

The catalytic activity of hen egg white lysozyme at varying levels of hydration was measured by monitoring the rate of formation of products in the hydrolysis of N-acetylglucosamine [26]. Figure 1 shows the enzyme hydrolytic

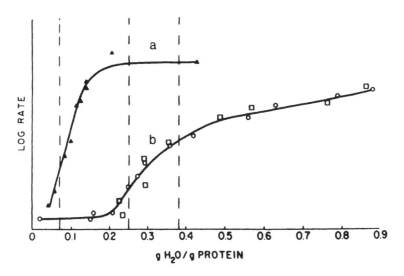

Figure 1   Dynamic properties of lysozyme as a function of hydration: (a) hydrogen exchange rate and (b) enzymatic activity (□) and rotational relaxation rate of the ESR probe Tempone (o). The onset of enzyme activity is strongly correlated with the dynamic properties of the protein matrix, measured by hydrogen exchange and ESR. These effects also appear to correlate with several other time-averaged spectroscopic and thermodynamic properties [28]. (Reproduced with permission from Ref. 18.)

activity and the dynamics of an ESR spin label, Tempone, noncovalently bound to the enzyme as a function of hydration level. The strong correlation between activity and increased motion of the spin label suggests that a degree of flexibility is required before lysozyme can exhibit activity. It further suggests that the necessary flexibility arises from the water. This correlation has been widely used to explain the strong dependence of catalytic activity on water content for enzymes suspended in nearly anhydrous media.

It should be noted that the hydration level of 0.20 h (g H_2O/g protein), which defines the onset of enzyme activity, is significantly less than that needed for monolayer coverage, the latter corresponding to about 0.46 h [29]. Hydration studies of porcine pancreatic lipase [30], bovine α-lactalbumin [31], and ribonuclease A [21] also show that activity may commence at hydration levels well below monolayer coverage.

The spectroscopic studies provide some explanation concerning the molecular role of water in inducing conformational change and increasing flexibility. When these studies are taken together, several conclusions can be drawn.

1.   The first hydration processes (hydration levels up to 0.10 h) are proton redistribution to the acidic and basic groups, followed by solvation of charged (and some exposed polar) groups.
2.   The increase in mobility that results from hydration allows solvent access to internal peptide groups. With access to the protein interior, water can compete with the peptides for hydrogen bonding, which may result in conformational changes. Conformational changes are complete at or before 0.20 h (i.e., before the onset of enzyme activity).
3.   Catalytic activity sets in well before hydration of nonpolar surface residues.

The sequential changes that occur upon hydration of lysozyme are summarized in Table 1.

Protein Hydration: Structural Aspects

Careri and coworkers [28] have correlated data on the hydration of lysozyme powders in an effort to obtain a unified picture of the hydration process. The formation of carboxylate ions at low levels of hydration (<0.05 h) is evidenced by the increased intensity of the IR absorbance at 1580 cm[-1]. This implies that water effects deprotonation of the carboxylic acid groups (presumably accompanied by protonation of basic groups) to given the protein ionization state found in solution. Proton redistribution is complete by 0.07 h. At very low hydration levels (0.07 h), water molecules are tightly attached to the protein via

Table 1.   Events Accompanying the Sequential Hydration of Lysozyme [19]

Hydration (g water/ g protein)	Water events	Protein event
0.0	Charged group hydration Polar group hydration	Proton redistribution (normal pK) Protein loosens up
0.1	Acids saturate Side-chain polar saturation	(Side-chain and backbone conformational shifts)
0.2	Peptide NH saturates	Enzyme activity returns Dynamic increase to solution value with increase in activity
0.3	Peptide CO saturates Polar group monolayer coverage Apolar surface coverage	
0.4	Complete monolayer coverage	

two or more hydrogen bonds and experience restrictions on free rotation. Above 0.07 h, waters are apparently less tightly bound and are more mobile [32].

The further addition of water from 0.1 to 0.2h represents hydration at amide, carboxylate, and other charged or polar sites [28]. At 0.20–0.25 h, about half the quantity of water as that required for complete coverage of the protein surface, the carboxylate and carbonyl sites are saturated. This hydration level corresponds to the point at which enzymatic activity is first observed (Fig 1).

Spectroscopic and thermal properties are insensitive to hydration at levels exceeding 0.38 h. The heat capacity reaches a constant value equal to that in dilute solution, which is believed to represent addition of bulk water to the system. The disagreement between the quantity of water calculated for monolayer coverage (0.46 h) and the value that corresponds to saturation in the heat capacity (0.38 h) can be interpreted in two ways. According to Rupley et al. [26], the water molecules in the vicinity of the protein are locally ordered and occupy greater areas per water molecule (20 $\mathring{A}^2$) on average than bulk water (10 $\mathring{A}^2$). Therefore, calculations of hydration levels for monolayer coverage based on the area occupied by bulk water overestimate the true quantity of water associated with the protein surface. On the other hand, Finney and Poole [19] suggest that at hydration levels exceeding about 0.35 h, polar group coverage is complete and is followed by coverage of exposed nonpolar groups by waters (complete at 0.45–0.50 h) that bridge water molecules already interacting with polar groups.

From Raman spectra of lysozyme glasses, Poole and Finney [23] concluded that any hydration-induced changes in the protein conformation are probably small and are complete at about 0.20 h. According to Rupley [18,28], no significant conformational changes accompany lysozyme hydration. In contrast, in a nuclear magnetic resonance study of two [113]Cd-substituted metalloproteins, parvalbumin and concanavalin A, Marchetti and coworkers [33] observed that the line widths of the [113]Cd magic angle spinning spectra showed a considerable degree of local disorder at the metal-binding region of both proteins in the lyophilized powder. Hydration of these proteins resulted in the reduction of the line width by a factor of 4, indicating a restoration of some local structure upon interaction with water.

Protein Hydration: Dynamic Aspects

The ESR experiments of Rupley et al. [26] described earlier (Fig. 1) demonstrate a significant change in the motion of a surface probe between hydration levels of 0.20 and 0.25 h. The increased mobility persists at hydration levels corresponding to complete dissolution of the protein. The abrupt change in mobility at 0.20 h, which correlates with the onset of enzyme activity, has led to the belief that at a certain degree of flexibility is needed before the enzyme can exhibit activity.

The hydration-induced flexibility changes observed in lysozyme motivated Poole and Finney [23] to perform NMR measurements of the exchangeability of amide protons as a function of hydration. The exchange method is a well-established probe of molecular flexibility [34–36]. The exchange rates of specific protons reflect their local environment and, more importantly, solvent accessibility and participation in hydrogen-bonded structures. Figure 2 shows the residual deuterated amide population of deuterated lysozyme as hydration is increased [23]. An initially steep decrease in the remaining deuterated amides is followed by a leveling off at about 0.05 h. There is a continued decrease in the deuterated amide population as hydration is further increased. These ob-

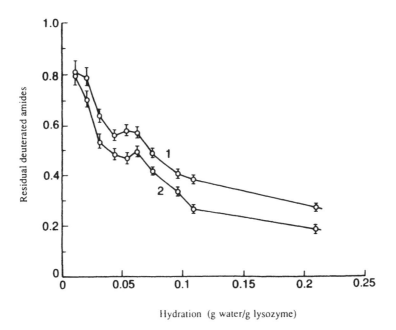

Figure 2 NMR measurements of the rate of hydrogen exchange as a function of hydration in lysozyme. The hydrogen exchange data have been interpreted as a hydration-induced change in the mobility of the protein matrix. The two curves correspond to different methods of determining the residual deuterated amides. (Reproduced with permission from Ref. 19).

servations suggest that the molecule is relatively rigid at low humidity, and therefore only the easily accessible surface deuterons can exchange with solvent protons. Previously inaccessible amides are able to exchange only at hydration levels exceeding 0.07 h. The increased amide accessibility implies solvent penetration or local unfolding motions, and therefore greater flexibility.

Parak [20] measured the mean-squared displacement of iron in [57]Fe-labeled myoglobin using Mössbauer spectroscopy, with results that also indicate that the internal mobility of the molecule increases with the level of hydration. In addition, a correlation was observed between protein mobility and the mobility of the surrounding water, showing that water of hydration is bound to the protein strongly enough to be affected by the protein motions.

What is the role of hydration in effecting this increased protein mobility? The hydration level at which lysozyme activity is first noted (0.20 h) corresponds to complete hydration of charged residues and is followed by partial polar group coverage and solvent penetration to the interior parts of the protein (Table 1). A consequence of solvent penetration is that water competes with the peptides for internal hydrogen bonding. Since hydrogen bonds between water and peptides are roughly of the same energies as hydration bonds internal to the protein, the ensuing dynamic equilibrium results in increased protein flexibility. It has been proposed that water is acting essentially as a plasticizer to increase protein flexibility [32].

The statement that water confers "flexibility" to the protein does not specify what "flexible" means in terms of molecular motions. Theoretical studies of internal dynamics of proteins have provided some insight into the nature of motions of protein atoms about their equilibrium positions [37,38]. Some attempts have been made to incorporate solvent effects in molecular dynamics calculations by considering the dynamics of the protein and the surface-bound van der Waals solvent as a whole. Molecular dynamics simulations of BPTI (bovine pancreatic trypsin inhibitor) in vacuo and with van der Waals solvent [39] suggest that atomic fluctuations in the protein can be described in terms of local oscillations superimposed on collective motions. The former have picosecond times scales and tend to be uniform over the protein structure. The collective motions, which can involve the cooperative motions of groups of atoms, have longer times scales (1–10 ps). It is these motions that can be translated to different parts of the protein and are believed to be most sensitive to the solvent and other environmental factors. More recently, a molecular dynamics simulation of ribonuclease T1 including solvent and hydrogen bonding effects, indicated that the solvent increases collective motions both at the surface and at the interior of the protein [40]. The increased fluctuations were attributed to the breakage of protein–protein hydrogen bonds and re-formation of protein water hydrogen bonds.

If hydrogen bonding is ultimately responsible for the flexibility needed for enzymatic activity, water appears to be unique in that it forms hydrogen bonds of "just the right strength" (i.e., comparable to hydrogen bonds within the protein). It has been noted that D_2O, in spite of its apparent similarity to H_2O, behaves very differently from water in terms of conferring flexibility to ribonuclease A [21]. The difference in behavior between H_2O and D_2O toward hydrating the protein must be related to the greater energy of the O—D bond, which attenuates the tendency of D_2O to participate in deuterium bonding.

PROTEIN STABILITY IN SOLUTION

The stability of a protein in solution is measured by the difference in the Gibbs free energy between the correctly folded and unfolded protein. The free energy of stabilization for a folded protein, or the free energy change accompanying the unfolding reaction F→ U, is usually presented as the sum of free energy contributions from various sources [41]:

$$\Delta G_{\text{F-U}} = \sum \Delta g_{i,\text{int}} + \Delta g_{i,s} + \Delta G_{\text{el}} + \Delta G_{\text{conf}} \tag{1}$$

The $\Delta g_{i,\text{int}}$ are the individual contributions to stabilization through short-range interactions such as hydrogen bonds, van der Waals forces, and salt bridges, while the $\Delta g_{i,s}$ terms represent short-range interactions with solvent. ΔG_{el} is the free energy change associated with long-range electrostatic interactions between charges and helix dipoles, etc. ΔG_{conf} accounts for the configurational entropy loss in the folded polypeptide chain. Factors that lower the free energy of the folded form and/or raise that of the unfolded forms will increase stability. Though many environmental factors (temperature, pH, salt concentration, etc) are known to affect stability, our concern here is primarily with the solvent.

The Hydrophobic Interaction

The affinity of different amino acid side chains for water has been considered to be one of the main factors governing the structures of proteins in solution [42]. For this reason, there has been sustained interest in the solvation properties of amino acid side chains [43,44]. To quantify the different affinities of amino acid side chains for water, many scales of hydrophobicity have been proposed [43-48]. These scales, derived from measurements of the distribution coefficients of the amino acids between an aqueous phase and a nonaqueous one [43,44,46,48] or a surface [47] correlate reasonably well with the distribution of amino acids between the surface and interior of globular proteins.*

Matthews and coworkers [49] studied the effects of hydrophobic interactions on the stability of phage T4 lysozyme in which Ile 3 was replaced by 13 different amino acid residues. Ile 3 is largely buried in the major hydrophobic core of the C-terminal lobe and contacts four other buried hydrophobic side chains. The incremental free energy of stabilization of the 13 variants was found to correlate with the hydrophobicity of the side chain, as measured by transfer free energies from water to ethanol. The majority of substitutions were less hydrophobic than Ile and were therefore destablizing. A similar dependence of protein stability on the hydrophobicity of a single substituted residue was observed by Yutani and coworkers [13] in the α subunit of tryptophan synthetase and by Fersht and coworkers [51,52] in barnase.

Several groups have attempted to increase protein stability via hydrophobic interactions [14,53,54]. However, side chain solvent affinities cannot be easily changed without affecting the packing of the protein interior and the associated van der Waals interactions. In reality, hydrophobic interactions must be considered to be an integral combination of van der Waals interactions between nonpolar groups and hydration effects of those groups [55]. Changes in packing can lead to structural distortion and strain. Matthews and coworkers [53] identified cavities in phage T4 lysozyme that might be able to accept hydrophobic residues more bulky than those in place in the wild-type enzyme. For the two mutations studied, the resulting enzymes were less stable than the wild type. The increase in stabilization that might be achieved by the increased van der Waals contacts and hydrophobic contributions of a bulkier residue was offset by local structural perturbations and strain [53,56]. These observations reflect the unusually efficient packaging of the protein interior. The packing density for many different globular proteins is 0.75 [55]. This value is closer to the value for crystals of small organic molecules than to organic liquids, which are close to 0.44. According to Karpusas et al. [53], increasing hydrophobic and van der Waals interactions by filling cavities with bulkier hydrophobic residues is unlikely to be a generally useful strategy for engineering thermostable enzymes.

---

*The assumption underlying such a comparison is that the interior of the protein can be modeled by a nonaqueous solvent. Although the transfer free energies can provide a good qualitative measure of the preference of the amino acid side chain for either the solvent or the protein interior, there are doubts about the quantitative utility of these measurements because the protein interior is much more complicated than can be represented by any one solvent [44,50]. See discussion below.

We can estimate solvophobic effects for substituting amino acids in a protein dissolved in a nonaqueous solvent with an approach that is analogous to the approach used in the aqueous studies described above. Protein stability is assumed to be related to the transfer free energies from the protein interior, as modeled by ethanol, to a nonaqueous solvent. The free energy of transfer of a residue from the protein interior (ethanol) to solvent s, ΔG_{e-s}, is imply the difference between the two transfer free energies for ethanol–water, ΔG_{e-w}, and solvent–water, ΔG_{s-w}:

$$\Delta G_{e-s} = \Delta G_{e-w} - \Delta G_{s-w} \qquad (2)$$

From data on transfer free energies from ethanol and N-methylacetamide (NMA) to water at 37 C [57], ΔG_{e-NMA} can be calculated for the 20 amino acid side chains. In Figure 3, ΔG_{e-NMA} is plotted against the virtual free energies associated with the observed distributions of amino acids between buried and solvent-accessible positions, $\Delta G_{bur-acc}$, calculated by Radzicka and Wolfenden [44] from the relative accessibilities of amino acids in 12 proteins [58]. Except for the polar side chains of arginine, lysine, glutamine, and histidine, the two sets of data show good correlation. These results indicate that the transfer of amino acid side chains from ethanol to NMA is directly related to the degree to which the different residues prefer to reside within the interior of a protein. In other words, the relative ordering of side chains on the solvophobicity scale is largely unchanged when NMA is substituted for water. However, the values of ΔG_{e-NMA} are very near to zero; the values of ΔG_{e-NMA} compared to ΔG_{e-w} indicate that the solvophobic effect in NMA is small compared to water. This qualitative result is insensitive to the choice of solvent. If the hydrophobic effect, as measured solely by side chain transfer free energies, is the dominant factor determining protein stability in an aqueous environment, then dissolution of the protein in a nonaqueous one would result in significant destabilization. However, membrane proteins and certain proteins that are soluble in polar nonaqueous solvents (e.g., the seed protein crambin) do not have interiors that are significantly different from water-soluble globular proteins with respect to amino acid composition [59]. This provides a strong argument that the hydrophobic effect may be less important than is often assumed. To the extent that van der Waals interactions, which are not intrinsically sensitive to the solvent, and hydrogen bonding play a role in what appears to be hydrophobic stabilization [55], the destabilizing effects of replacing water with another solvent may not be realized. In other

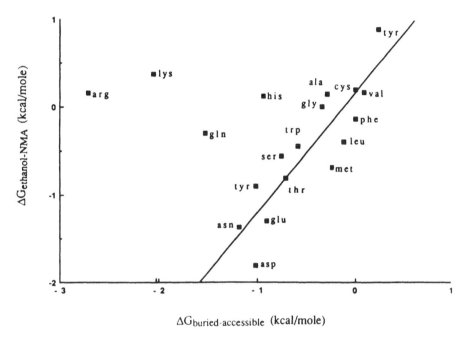

Figure 3 Transfer free energy of the amino acid side chains from ethanol to NMA plotted against the virtual free energy of transfer from the interior of the protein to the surface (data from Radzicka and Wolfenden [44]) at 37°C. If ethanol is a reasonable model of the protein interior, these results indicate that the preference of the amino acid side chain to reside on the surface or the interior would not change radically upon transferring a protein from water to NMA. However, the magnitude of the solvophobic contribution to protein stability would be greatly reduced.

words, the interiors of water-soluble proteins may not differ greatly from the optimal interior for a protein stable in nonaqueous solvents.

Hydrogen Bonding

Hydrogen bonding makes significant contributions to formation of secondary structure and stability, in addition to placing tight constraints on the conformations of folded proteins. A fully extended polypeptide chain is able to satisfy its hydrogen bonding potential through interactions with solvent. Similarly, almost all the polar groups in folded proteins can satisfy their hydrogen bonding requirements through a combination of internal protein–protein hydrogen bonds and bonds with the solvent. For RNase-S,7% of main chain amide groups and 15% of the carbonyl oxygens are unable to satisfy their hydrogen bonding requirements, while only 3% of the polar groups in lysozyme are not involved in hydrogen bonding [60]. This tendency to maximize hydrogen bonding limits the possible folded protein conformations.

The importance of the hydrogen bonding interaction to protein stability can be seen from a study of the thermostability of 13 different variants of T4 lysozyme containing amino acid substitutions for Thr 157 [61]. In the wild-type lysozyme, Thr 157 forms a hydrogen bond with the buried main chain of Asp 159. Amino acid substitutions that resulted in the most stable mutant enzymes were the ones able to form a hydrogen bond to Asp 159. Substitution of Thr 157 by Ile had been shown earlier to destabilize the phage lysozyme [62]. Stability studies on the 13 variants indicate that the loss of the hydrogen bond dominates the effects on the stability of the folded protein and further supports the conclusion of Grutter and coworkers [62] that the destabilizing influence of the Ile substitution was due to the loss of the Thr 157-Asp 159 hydrogen bond.

Fersht [63] has argued that mutations leaving unpaired, uncharged hydrogen bond donors or acceptors decreases binding energy by 0.5-1.8 kcal/mol. When charged species are left unpaired, the effect on binding energy is significantly larger, (3- 6 kcal/mol), reflecting the higher bond dissociation energies for hydrogen bonds to charged groups. Considering that folded proteins are only marginally stable (5-20 kcal/mol), a single unsatisfied hydrogen in the folded state can result in the loss of a significant fraction of the stability of the folded molecule [60,64].

The introduction of new or improved hydrogen bonds is expected to increase the stability of the folded protein. The replacement of Asn 218 by Ser in subtilisin BPN′ improves bond parameters for several existing hydrogen bonds and stabilizes the enzyme by roughly 1 kcal/mol at 58.5 C [65]. Other substitutions at the same position resulted in relatively less stabilization, and the rankings were consistent with simple hydrogen bonding considerations [65].

In aqueous solution, water competes for the hydrogen bonding sites in the protein, forming hydrogen bonds of strength similar to the internal protein-protein hydrogen bonds that are broken. The resulting dynamic equilibrium between conformational substates may be the cause of the increased flexibility observed upon hydration of dry proteins (see above, "Protein Hydration: Dynamic Aspects"). The formation and breakage of hydrogen bonds between the solvent and the protein can be described in terms of the following equilibrium:

$$p-s + p-s = p-p + s-s \qquad (3)$$

where p–p, s–s, and p–s refer to hydrogen bonds between protein and protein, solvent and solvent, and protein and solvent respectively. The equilibrium above can be considered to be the sum of three equilibria involving the rupture and synthesis of hydrogen bonds as follows:

$$p + p = p-p$$

$$s + s = s-s \qquad (4)$$

$$p-s = p + s$$

The total change in Gibbs free energy for the reaction of Equation (3), $\Delta g_{i,\text{H-bond}}$, is a weighted sum of the changes in free energies involved in the three substeps. Water forms hydrogen bonds with the protein that are similar in enthalpy to protein–protein and water–water hydrogen bonds. Since the total number of hydrogen bonds is conserved in the reaction, the net process described above is essentially isoenthalpic. The main contribution to the free energy therefore arises through entropy effects from the release of enzyme-bound water [64].

The effect of solvent on hydrogen bonding equilibria has been studied using N-methylacetamide as a model for a peptide group forming a hydrogen bond with another peptide [66]. The equilibrium constant for the dimerization of NMA was determined in three solvents: carbon tetrachloride, dioxane, and water. Since the first two solvents do not form hydrogen bonds or form only weak bonds, the enthalpy of the foregoing reaction is nonzero (–4.2 kcal/mol in CCl₄ and –0.8 kcal/mol in dioxane). In water, on the other hand, the reaction is isoenthalpic. The equilibrium constants at 25° C in the three cases are 5.2, 0.55 and 0.005 M⁻¹, respectively, showing that the monomeric state is strongly favored over the dimer in aqueous solution. This reflects, to an extent, the strong competitiveness on the part of the water for forming hydrogen bonds with amide groups.*A recent theoretical study of hydrogen bond formation between two molecules of formamide in water and carbon tetrachloride concluded that

---

*Because water was in excess over NMA it is impossible to conclude from this study whether NMA prefers to form hydrogen bonds with water or with itself.

the hydrogen bond is approximately 8 kcal/mol more stable in the nonpolar solvent than in water [67].

In a solvent that forms weaker hydrogen bonds than water, or no hydrogen bonds at all, the hydrogen bonding contribution to the free energy of protein folding could become very important. The result of removing protein–solvent hydrogen bonds would be to increase the free energies of both folded and unfolded states. However, in general, in a properly folded protein there will be more internal hydrogen bonds satisfied than in unfolded or misfolded forms. Thus in a nonaqueous solvent, protein–protein hydrogen bonding can mean a net increase in the number or quality of hydrogen bonds upon folding, and therefore may become a particularly important determinant of protein stability. On the other hand, solvent-accessible polar groups, which had been hydrogen bonded to solvent in the aqueous environment and would remain unsatisfied in the nonaqueous one, may tend to be destabilizing. This is discussed further under "Hydration," below.

Electrostatic Interactions

Electrostatic interactions leading to formation of ion pairs contribute to the stability of folded proteins. Although ion pair formation has been shown to increase the stability of α-helices in short peptides [68], most salt bridges in proteins cross-link different elements of secondary structure and therefore serve to stabilize tertiary structure [69].

Protein stabilization has been accomplished through the introduction of salt bridges. However, the introduction of stabilizing salt bridges cannot be accomplished without careful consideration of the local protein and solvent environment. Proteins stabilize ion pairs by providing a relatively fixed complementary electrostatic environment of charges or permanent dipoles (hydrogen bonds, fixed waters, etc.) [70]. Similarly, the environment can serve to destabilize an ion pair, if the appropriate complementary interactions are not available or the polarization is unfavorable. An ion pair in a hydrophobic environment is unstable; protein interiors surrounding ion pairs are polar [71].

Other electrostatic contributions to protein stability include charge interaction with helix dipoles [72] and metal ion binding. Binding of metal ions is known to increase enzyme stability for a number of thermostable enzymes including thermolysin [73]. Pantoliano and coworkers [54,74] evaluated the effects of engineering a higher affinity for Ca^{2+} into subtilisin BPN′. The substitution of Gly and Pro with Asp in the vicinity of a low affinity Ca^{2+} binding site resulted in an increase in the affinity for Ca^{2+} and was accompanied by a shift in the Ca^{2+} concentration dependence of the half-life for thermal inactivation. However, a relatively small change in the maximum enzyme half-life at high calcium concentrations was observed ($t_{1/2}$ increased from 252 minutes for the wild-type enzyme at 65 C to 272 minutes for the variant exhibiting the highest Ca^{2+} affinity). The effects of individual mutations on the electrostatic potential in the vicinity of the Ca^{2+} binding site were additive.

The effects on electrostatic interactions of transfer to a nonaqueous solvent are difficult to predict. If the nonaqueous solvent has a low dielectric constant, surface ion pairs may be significantly destabilized. Hydration stabilizes both free charges and ion pairs. Desolvation of an ion pair may result in conformational changes that cause reorientation of protein dipoles to stabilize the charges. Substitution of water by nonaqueous solvents often promotes metal binding, since water is a reasonably good ligand for many metals. Without competition from water, metal binding may enhance stability in nonaqueous solvents.

Hydration

Electrostatic solvation energies of between –25 and –50 kcal/mol have been calculated by Gilson and Honig [75] for individual charged surface atoms in rhodanese and crambin. The magnitude of the hydration energy of charged residues places a large penalty on burying these residues in the folded structure, resulting in the observation that charged groups are usually located at or near the protein surface [69,76]. When charged groups are found in the protein interior, they are usually stabilized by salt bridges and hydrogen bonds, which compensate for the desolvation. Fully buried charged residues are often functionally important.

The importance of the solvation of charged residues in the folded state was demonstrated in a comparison of the solvation energies of misfolded and crystal structures of *Themiste dyscritum* hemerythrin and the VL domain of an antibody [75]. Because the folded proteins optimize hydration of charged residues, the electrostatic free energy contribution form solvation strongly favors the crystal structures over misfolded structures. For the same extent of solvent exposure, the amino acid residues with the strongest solvent interactions are those with ionizable side chains. The main chain nitrogen atom at the N-terminus is also generally exposed to solvent and has a large electrostatic free energy of solvation.

Transfer of the protein from an aqueous milieu to a nonaqueous one that is a poorer solvator of charged and polar residues would have a large, negative effect on the solvation contribution to protein stability. To recover some of the lost solvation energy, charges interact with the surface charges on other protein molecules, leading to aggregation and drastic changes in solubility. Accompanying this effect is the possibility of conformational changes in the protein to allow neighboring surface charges to form salt bridges or hydrogen bonds with other residues. In addition, the pK_a values of solvent-exposed charged residues will change, which can influence protein conformation and catalytic activity.

Configurational Entropy of the Unfolded Protein

Stabilization of the protein can be achieved by either increasing the stability of the folded form or destabilizing the unfolded form(s). The introduction of disulfide linkages decreases the configurational entropy of the unfolded state, thereby decreasing the entropy to be gained in the unfolding reaction. A disulfide bond engineered into dihydrofolate reductase increases resistance to denaturation by guanidine hydrochloride but not thermal stability [77]. Subtilisin BPN' containing an engineered disulfide bond is significantly more stable with respect to thermal inactivation and inactivation in 8 M urea [12]. Disulfide bonds introduced into phage T4 lysozyme at two positions increase the enzyme's melting temperature by 6.4 and 11.0 C [15]. A disulfide bond identified using computer search techniques [78] and engineered between subunits of the N-terminal operator binding domain of λ repressor increases the stability of the dimer with respect to thermal denaturation and denaturation with urea [79]. The presence of the disulfide, which forms spontaneously, increases the affinity of the dimer for operator DNA.

The success of disulfide engineering depends on the effects of cysteine substitutions in the folded form as well as the unfolded state(s). In several cases, the net result of adding a new disulfide bridge has been a decrease in stability [15,80]. Unfavorable nonbonded interactions and strain induced in the folded form by the introduction of disulfides can obliterate any entropic contribution to stabilization [81]. Matsumura and coworkers [15] found that the largest benefits were obtained for disulfides with large loop sizes, placed in flexible parts of the molecule. The flexibility appears to allow the disulfide bridges of favorable geometry to form with a minimum of strain [15,82].

The configurational entropy of the unfolded protein can also be reduced by making substitutions of the type Gly$\rightarrow$ Xaa or Xaa or Xaa$\rightarrow$ Pro (Xaa denotes any amino acid), which reduce the degrees of freedom available in the unfolded state [83]. The substitution Gly 77 $\rightarrow$ Ala increased the free energy of unfolding of phage T4 lysozyme by 0.4 kcal/mol at 65 C and pH 6.5, while Ala 82$\rightarrow$ Pro increased it by 0.8 kcal/mol [83]. It was noted that an alternative explanation for the stabilizing influence of the Gly$\rightarrow$ Ala mutation is the increase in helical propensity afforded by alanine. A Gly$\rightarrow$ Ala substitution in an α-helix of λ repressor [84] also increases thermal stability, but the effect was attributed to the replacement of Gly, a poor helix former, with a residue that is a much better helix former. The stabilization that resulted from a Gly→ Ala replacement in *B. stearothermophilus* neutral protease was attributed to a combination of hydrophobic effects and helix stabilization [14]. As noted by Matthews et al. [83], the entropic benefit of the Gly→ Ala mutation should be available regardless of whether the Gly is located in an α-helix.

Similar entropic considerations should be valid for proteins in nonaqueous solvents, if unfolded states are accessible. The effect on entropic contributions of transferring a protein to a nonaqueous solvent would then depend on the nature of the unfolded states available to the wild-type and engineered enzymes in that solvent. An interesting benefit of disulfide bridges is that they may guard against aggregation in the unfolded state by limiting the possible unfolded conformations, thereby reducing the rate of irreversible inactivation [81,85]. This effect may also be important in nonaqueous solvents.

Additive Effects of Mutations on Protein Stability

The individual Δg_i terms in Equation (1), representing local contributions to the stabilization of the folded protein, are summed in the total free energy of unfolding. The structural perturbations that result from amino acid substitutions are often accommodated with only localized changes in the protein and solvent [61]. This means that the effects of conservative changes in the amino acid sequence in spatially distant parts of the protein can be additive. Additive effects on protein thermostability have indeed been observed in a number of cases [16,54,84,86]. Pantoliano and coworkers [54] found six different amino acid substitutions that stabilize the folded form of subtilisin BPN', with increases in the free energy of unfolding ranging from 0.3 to 1.3 kcal/mol for each substitution. When the individual mutations were combined, they resulted in a very nearly additive increase in the free energy of unfolding of 3.8 kcal/mol at 58.5 C. This was accompanied by a 300-fold reduction in the rate of the protease's irreversible thermal inactivation. Matsumura and coworkers have shown that increases in the melting temperature of phage T4 lysozyme containing two or three engineered disulfide bonds were approximately additive; the melting temperature of the mutant with three disulfides was 23.4 C higher than that of the wild-type enzyme [16].

These studies have shown that (1) enzymes can tolerate multiple mutations without substantial alterations in three-dimensional structure and (2) mutations that increase stability can be combined to obtain enzymes that will function under conditions significantly removed from the optimal conditions for the wild-type enzyme.

ENZYME ENGINEERING FOR IMPROVED
STABILITY IN NONAQUEOUS SOLVENTS

In engineering an enzyme for use in a nonaqueous solvent, the goal is to maximize its activity and increase its lifetime in the new environment. Polar solvents are useful for their ability to solubilize a wide range of substrates. In addition, it is often desirable to minimize unfavorable side reactions by reducing the water content to the minimum compatible with reasonable activity. For example, in using hydrolytic enzymes effectively to carry out their reverse synthetic reactions, one must limit hydrolysis of the product, and this is best done by removing as much water as possible from the system. Therefore, in this discussion of engineering nonaqueous solvent-stable enzymes, it is assumed that the goal is to stabilize the enzyme in a polar organic solvent and to minimize its dependence on water.

Considerable evidence that enzymes can be engineered in a rational manner to improve their stability in aqueous media has accumulated. Significant improvements in thermostability in aqueous solution have been achieved, and the stabilizing results of mutations have in many cases been shown to be additive. At least some of the types of mutations that improve thermostability are also likely to improve stability in the presence of denaturants such as organic solvents.

It is important to note that in this chapter we are not considering specific effects of the solvent on catalysis. These effects are obviously extremely important in determining enzyme activity in nonaqueous media, but they are peculiar to each enzyme–substrate pair and catalytic mechanism. Instead we are focusing solely on the more general problem of protein stability. To improve protein stability, the goal of any modification in the protein sequence is the stabilization of the active form of the enzyme in the new solvent environment relative to other, inactive forms that are accessible in that environment. In other words, the reaction that must be considered in choosing the appropriate amino acid substitutions is no longer $F \rightarrow U$ in aqueous solution, but rather $F \rightarrow I$ in solvent, where I is any inactive form, folded or not. This difference in the reaction is important for two reasons.

1. The enzyme may not dissolve in the nonaqueous solvent and therefore may not be able to undergo transitions from folded to unfolded forms. The inactivated enzyme may be very similar in structure to the active form. However, the solvent (or lack of solvent!) causes inactivation by other means: inflexibility, subtle conformational changes, changes in ionization state, etc.

2. If the enzyme is dissolved in the nonaqueous solvent (or water–solvent mixture), unfolding is now possible. The forces that contribute to stabilization of the active, properly folded form will attain a new balance that depends on the solvent. Therefore rules for designing proteins for stability in nonaqueous media must take into account the solvent effects on interactions in all possible conformations.

Enzymes Suspended in Nonaqueous Solvents

Most enzymes are insoluble in organic solvents with low water contents and therefore remain suspended in powder or particle form. Enzymes do not all respond in the same way to suspension in nonaqueous solvents. Some are active only in nonpolar, water-immiscible solvents, while others retain some activity in polar, water-miscible solvents. For example, horse liver alcohol dehydrogenase is catalytically active to some extent in a wide range of solvents, but yeast alcohol oxidase is active only in water-immiscible ones [7]. Subtilisin is weakly active in a transesterification reaction in dry dimethylformamide, whereas chymotrypsin shows no detectable activity under similar conditions [87]. Both subtilisin and chymotrypsin are active in dry octane.

The variation in sensitivity to organic solvents exhibited by different enzymes has been attributed to several different protein properties. One possible contributor to this behavior is conformational stability: high stability might limit the possible effects of lyophilization on conformation and activity [87]. Structural heterogeneities observed in dehydrated metalloproteins [33] and conformational changes during hydration of lysozyme [23] were discussed in the section on protein hydration. However, similar conformational changes have not been detected for enzymes suspended in nonaqueous solvents. ESR studies of the active site of lactate dehydrogenase suspended in various solvents containing 0.1 to 10% water yielded no evidence of conformational change upon addition of water, although dramatic increases in enzyme activity were observed at the higher water contents [88]. Similarly, no conformation changes were apparent in a solid-state NMR study of α-lytic protease suspended in anhydrous octane and acetone [89]. The ^{15}N isotropic chemical shifts of α-lytic protease suspended in anhydrous organic solvents showed no difference in the tautomeric structure and hydrogen bonding of His 57 in the catalytic triad in the organic solvents versus water. The large difference between the catalytic efficiencies of α-lytic protease in octane and acetone (k_{cat}/K_m = 5400 and 3.3 M^{-1} m^{-1}, respectively) was attributed to other factors related to substrate binding or removal of water from the enzyme surface by the water-miscible acetone.

An explanation set forth for the increase in enzyme activity commonly observed with increased water content for enzymes in nonaqueous solvents at low water contents is an increase in protein flexibility, which is somehow critical for catalytic activity. This flexibility increase has been attributed to the "plasticizing" effect of water discussed in the earlier section on protein hydration [7]. Indirect evidence for the role of water in increasing flexibility in nonaqueous solvents has been reported for polyphenol oxidase in hexyl acetate [7]. However, no direct spectroscopic observations of increases in flexibility under these conditions have been reported. Nor does the flexibility argument explain why one enzyme is more active in nonaqueous solvents than another. In fact, it appears to contradict the hypothesis that conformational stability is the important parameter; increased conformational stability is often accompanied by decreased flexibility, as observed by hydrogen exchange [35] and other measurements [90]. Clearly, further studies of enzyme structure and dynamics in nonaqueous solvents are needed to clarify these issues.

For enzymes suspended in organic solvents, Klibanov and coworkers have proposed that the primary effect of polar nonaqueous solvents on activity is not through direct interactions of the solvent with the enzyme, but rather through the ability of water-miscible solvents to partition water away from the protein surface [7]. The ability of a particular solvent to dehydrate a protein can partially explain the variation among solvents in their ability to inactivate enzymes, but it does not elucidate the molecular mechanisms directly involved in inactivation. In the absence of a more complete understanding of the mechanisms involved, it is difficult to determine which types of amino acid substitutions will lead to improved performance of enzymes suspended in polar solvents. However, using information obtained from protein hydration studies and observations on the behavior of enzymes suspended in nonaqueous media, some reasonable strategies for enzyme stabilization through protein engineering can be proposed.

The transition to a fully unfolded state is kinetically inaccessible for the suspended enzymes, at least at moderate temperatures and low water contents. This inability to undergo unfolding transitions explains the high stability with respect to irreversible inactivation of enzymes suspended in dry organic solvents [91]. If unfolding is inaccessible or, at least, a very rare event, it may not be necessary at first to consider the details of all the forces that contribute to stability in solution (Eq. 1). Instead, it is reasonable to emphasize mutations that will tend to decrease the concentration of inactive forms of the enzyme by optimizing interactions between the folded, active state and the solvent. In other words, a strategy for improving enzyme stability is to direct mutations to those features that have been adversely affected by suspension in the solvent. In this scenario, emphasis would be placed on avoiding unfavorable interactions with the solvent or unfavorable effects of dehydration in the folded protein.

Observations from protein hydration studies may aid us in proposing specific mechanisms for improving enzyme performance in nonaqueous solvents. Unfortunately, a detailed view of the hydration process is not available for many proteins, and, therefore, we know little about how these processes depend on a protein's structure, amino acid composition, and surface chemistry. Events accompanying the rehydration of dry lysozyme were summarized in Table 1. Lysozyme activity was observed to follow proton redistribution and hydration of charged residues. In addition, the bulk of the changes in spectroscopic properties occur at low hydration levels corresponding to saturation of solvation at polar sites. These earlier hydration events are accompanied by loosening of the protein structure and, perhaps, small conformational rearrangements. At the higher levels of hydration, which correspond to nonpolar surface coverage, smaller, gradual changes in protein properties, including flexibility and enzymatic activity, are evident.

As a first approach to engineering enzymes to function in low water environments, it is reasonable to assume that the processes that occur before the onset of catalytic activity (proton redistribution, hydration of charged residues) are critical. The processes that occur at higher hydration levels (amide hydration, nonpolar group hydration) may be influential in improving activity, but are not critical. One solution to improving enzyme performance may then be to reduce the amount of water required for proton redistribution and the hydration of charged sites by selectively eliminating charged residues from the protein surface. This appears to be a reasonable approach from several viewpoints. Charged residues are well solvated in aqueous solution, and complete dehydration would result in a large increase in the system free energy. As discussed previously, desolvation may be accompanied by conformational changes to allow charged residues to become solvated by interactions with other polar groups in the protein. In addition, pKa changes accompanying dehydration could influence catalytic activity.

The result of selectively removing surface charges is to render the protein surface more hydrophobic. This strategy is not surprising: membrane proteins, whose natural environments are hydrophobic, do precisely that. Although the hydrophobicities of the interiors of transmembrane proteins are not significantly different from the interiors of water-soluble proteins, their membrane-exposed surfaces are significantly more hydrophobic [59]. The removal of surface charged groups may decrease the amount of water required for activation of the enzyme. An appropriate strategy would be to replace charged residues not implicated in biological function with neutral ones, without disrupting existing favorable interactions or introducing new unfavorable nonbonded interactions in the folded protein. A prudent approach would be to target for substitution by less polar side chains those residues not conserved in homologous proteins. If homologous protein sequences are not available, the sites targeted for substitution would exclude those participating in intramolecular hydrogen bonding or other important nonbonded interactions. It should be noted that altering surface charges can influence the pH dependence of catalytic activity [92].

It is possible that ion pairs on the protein surface will be significantly destabilized in very low water content systems, as discussed in the preceding section. In addition to removing free charges, one might consider the removal of exposed ion pairs to improve the stability of an enzyme in nonaqueous solvents. However, the removal of ion pairs is likely to have a deleterious effect on conformational stability, and the net effect on stability in nonaqueous solvents is difficult to predict.

It is very likely that interactions in the protein interior influence enzyme activity in nonaqueous solvents and how an enzyme responds to the removal of water; increasing conformational stability should reduce the unfavorable effects of nonaqueous solvents on structure. Amino acid substitutions that might enhance conformational stability are discussed in the section that follows.

Enzymes Dissolved in Nonaqueous Solvents

Few natural enzymes are soluble in anhydrous organic solvents. However, in some cases, it may be desirable to work with soluble enzymes or even to engineer greater solubility into insoluble enzymes. For example, soluble enzymes are needed to digest insoluble polymeric substrates. In addition, soluble enzymes are not subject to mass transfer limitations in homogeneous reaction systems. What are the special structural features of a protein that is both soluble and stable in nonaqueous solvents, and how might these features be incorporated into a protein design strategy? A useful model for protein design for nonaqueous solvents is afforded by cambin, a protein that is both soluble and remarkably stable in polar organic solvents such as methanol, dimethylformamide, and glacial acetic acid [93,94]. Crambin, itself insoluble in water, has several water-soluble homologs with similar folded structures [95–97]. Crambin and its homologs offer a convenient demonstration of protein engineering for stability in nonaqueous solvents [98].

Although it consists of only 46 amino acids, the crambin backbone is cross-linked by three disulfide bridges. There is a considerable degree of secondary structure in this small protein: two α-helices, a short β-sheet and a β-turn [99]. The inner bend between the helical stem and β-sheet contains the bulk of the charged residues and polar side chains in a highly interconnected network. Not surprisingly, crambin's surface is largely hydrophobic. In a study of the surface charge densities of 32 proteins, crambin was found to have the lowest overall surface density (0.65 charged groups per 100Å^2, while the mean value for the 32 proteins was 1.42) [100].

The positions of water molecules in the crambin crystal are particularly enlightening [101]. In addition to mediating interactions between charged residues, water molecules are hydrogen bonded to the protein backbone, especially at turn regions where protein–protein bonds have not formed. At several positions, polar side chains fold back to form hydrogen bonds to the backbone, reducing the number of sites available for interaction with water. The combination of a high degree of ordered secondary structure and the use of hydrophilic side chain atoms to satisfy hydrogen bonds leaves few sites at which hydrogen bonding to solvent water is necessary. In response to the paucity of favorable sites for interaction with the hydrophobic protein surface, water molecules maximize their interactions with one another, forming hydrogen-bonded pentagonal arrays.

Crambin's sequence has been compared to those of its water-soluble homologs viscotoxin A3 and α_1-purothionin to determine which substitutions might contribute to the new properties of the homologs [98]. The three disulfide bonds in crambin are also present in the water-soluble proteins. Viscotoxin A3 and α_1-purothionin are highly basic, containing numerous exposed lysine and arginine residues that have been replaced by hydrophobic residues in crambin. The few charges remaining in crambin participate in salt bridges or multiple hydrogen bonds with other side chains or bridging water molecules. A common substitution is one that removes a hydrophilic atom from solvent-exposed side chains of the water-soluble proteins, such as Thr→ Ile or Asn→ Ser. These substitutions remove surface hydrogen bonding sites occupied by water molecules and add to the nonpolar character of crambin's surface.

These features of crambin led us to propose a general set of rules for designing proteins that are stable when dissolved in polar nonaqueous solvents [98]. These rules included three important features: (1) a maximum number of internally fulfilled hydrogen bonds (2) a relatively hydrophobic protein surface, and (3) internal cross-links (e.g., salt bridges, disulfide bonds, other nonbonded interactions). A somewhat expanded set of design rules, with specific suggestions for protein engineering, is presented in Table 2 and discussed in detail below. In addition to providing general guidelines for engineering stability in nonaqueous solvents, these design rules may prove useful in choosing natural enzymes appropriate for use in nonaqueous media and appropriate for protein engineering studies.

To accomplish the goal of rational enzyme stabilization in nonaqueous solvents, it should be possible to apply some of the concepts that have proven

Table 2 Design Rules for Engineering Nonaqueous Solvent-Stable Enzymes [a]

Conformational stability

Introduce internal cross-links

    increase van der Waals interactions; maintain or improve packing
    introduce disulfide bonds
    introduce new or improve existing hydrogen bonds and other electrostatic
      interactions (see below)

Maximize internally fulfilled hydrogen bonds with high degree of ordered secondary
    structure; use side chains to fulfill main chain hydrogen bonding sites

Electrostatic interactions

Compatibility of the surface with solvent; more hydrophobic surface

Remove surface charges
Remove surface hydrogen-bonding sites

---

[a]Amino acid alterations that address these issues should result in improved enzyme stability in polar organic solvents. It is assumed that amino acid substitutions are chosen to minimize disruption of function and unfavorable nonbonded interactions. See text for discussion.

useful in engineering thermostability. The interiors of proteins stable in nonaqueous solvents are not likely to differ significantly from those of proteins that are stable in water. Therefore, part of a reasonable strategy for designing enzymes for nonaqueous solvents is to consider mutations that have been shown to stabilize the folded protein in aqueous solution. However, the mutations that result in improved thermostability and those that lead to improved stability in nonaqueous solvents may not be a perfectly overlapping set. As discussed in the earlier section on protein stability in solution, not all the forces that contribute to conformational stability in aqueous solution contribute equally in a nonaqueous solvent.

For example, some components of hydrophobic interactions are not expected to contribute significantly to the stabilization of a folded protein in nonaqueous solvents. In the discussion on protein stability in solution, it was shown that the solvophobic driving force for association of amino acid residues in *N*-methylacetamide is small compared to water, and therefore the purely solvophobic driving force for protein folding is likely to be small in NMA. This result is not unique for the ethanol–NMA solvent system. If the solvophobic effects of amino acid substitutions are well described by transfer free energy data, as is often assumed for explanations of hydrophobic effects in aqueous solution (see preceding section), one would predict that the hydrophobic contribution to protein stability would be largely lost upon transfer to a nonaqueous solvent. However, it is clear that the transfer free energies do not adequately describe the effects of tight packing of the protein interior and van der Waals interactions between nonpolar groups, and therefore cannot account for protein stability [55]. Packing considerations and van der Waals interactions will continue to contribute to stability in nonaqueous solvents and will probably dominate the effects of nonpolar amino acid substitutions. The precise consequences with regard to protein engineering for nonaqueous solvents are difficult to predict, because one is not talking about an identifiable force, but about a combination of complex interactions. However, it is reasonable to assume that so-called hydrophobic stabilization, to the extent that amino acid substitutions improve van der Waals interactions and packing of the protein interior, will continue to yield some measure of stabilization in nonaqueous solvents.

The introduction of stabilizing internal cross-links in the form of new disulfide bridges, hydrogen bonds, and certain electrostatic interactions should lead to improved stability in nonaqueous solvents. Carefully designed disulfide bridges will reduce the conformational freedom of unfolded enzyme forms, reducing the population of unfolded or inactive states. In addition, disulfide cross-links may limit aggregation and irreversible inactivation, which generally accompany protein unfolding.

Hydrogen bonding should have a particularly important role in protein stability in nonaqueous solvents. In a nonaqueous solvent unable to form hydrogen bonds or forming only weak bonds, the loss of water as a source of hydrogen bonds will have a large impact on protein conformation and stability. The inability of the protein to form hydrogen bonds with the solvent could possibly lead to alternate conformations that satisfy more hydrogen bonds internally, a result of the high energetic cost of leaving hydrogen bonding potential unsatisfied. In addition to conformational changes, protein flexibility could be affected if the solvent cannot replace broken protein–protein bonds with bonds to itself. To maximize protein–protein hydrogen bonding in the folded protein, a strategy for stabilization in nonaqueous media might include the introduction of new hydrogen bonds or improvement of existing ones. In choosing an enzyme for use in nonaqueous media, one containing a large number of hydrogen bonds (a high percentage of ordered secondary structure) would be a good first choice. Since the solvent may not be capable of satisfying surface exposed hydrogen bonding sites, these may be removed or, possibly, satisfied through the addition of external hydrogen bonding agents [7].

The importance of electrostatic interactions in determining the conformation of the folded protein in aqueous solution has been discussed. On transferring a protein to a nonaqueous solvent, the loss of the solvation energy is potentially highly destabilizing, since the native protein conformation in water maximizes the solvation contribution. The desolvation effects for solvent-accessible charges discussed in the preceding section on enzymes suspended in nonaqueous media should apply equally to dissolved proteins, and therefore the removal of charged residues from the surface should enhance the stability of the folded enzyme. In addition to dehydration effects, changes in the dielectric constant of the solvent are expected to change the magnitude of electrostatic free energies of salt bridges or charge interactions with helix dipoles, especially near the protein surface. As discussed previously, salt bridges in hydrophobic environments are unstable. It is possible that nonaqueous solvents will destabilize surface ion pairs that rely upon water for solvation and stabilization. Ion pairs that are stabilized by protein dipoles rather than solvent, on the other hand, may be stabilizing elements in nonaqueous solvents.

An additional complication of transferring a protein to a nonaqueous solvent is a change in solubility and protein–protein aggregation. Charged residues, originally solvated in water, can partially replace solvation by interacting with polar residues on the other protein molecules. The total free energy change experienced on transferring the protein to an organic solvent may now be complicated by intermolecular interactions that lead to deactivation. Modification of the protein surface to increase its hydrophobicity and increasing conformational stability through the mechanisms mentioned in Table 2 should help to reduce aggregation in nonaqueous solvents.

One finds, surprisingly, little in the literature regarding the effects of sequence alterations on protein stability in nonaqueous solvents. Recently, however, Wong and coworkers have reported that the highly thermostable subtilisin BPN' variant containing six amino acid substitutions described by Pantoliano [54] is also more stable in anhydrous dimethylformamide (DMF) [102]. The net effect of the six substitutions is a 50-fold slower rate of inactivation in DMF. (In 50% DMF, however, the variant and wild-type enzymes have similar stabilities.) Unfortunately, the amino acid substitutions were not studied individually, and therefore the contributions of each mutation to enhancing stability in the organic solvent cannot be quantified. For the purposes of testing the design rules in Table 2, one would wish to know to what extent the individual substitutions in this combination variant contribute to stability in DMF and whether all the substitutions that improve stability in aqueous media also enhance stability in DMF.

Protein engineering studies carried out in this laboratory on the proteolytic enzymes subtilisin E and α-lytic protease have demonstrated that stability in

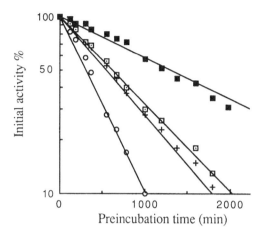

Figure 4   Stability of wild-type and mutant subtilisins E in 80% (v/v) DMF at 30 C. Co-solvent is 0.01 M Tris, pH 8.0, 2 mM CaCl₂. (o) wild-type; (+) Asp 248 → Asn; (▣) Asn 218 → Ser; (■) Asp 248 → Asn + Asn 218 → Ser.

nonaqueous solvents can be enhanced following the general procedures outlined in Table 2. As shown in Figure 4, substitution of a charged surface residue (Asp 248) by the uncharged amino acids Ala, Leu, and Asn increases the stability of subtilisin E in 80% DMF (Robinson, et al., in preparation). The mutation Asn 218 → Ser, known to increase the stability of subtilisin BPN′in aqueous media by improving hydrogen bonding [65], also enhances the stability of subtilisin E in 80% DMF. The effects of two mutations are cumulative: the double mutant Asp 248 → Asn + Asn 218 → Ser is more stable than either single mutant.

The substitution of charged residues on the surface of α-lytic protease also enhances the enzyme's stability in DMF. A systematic study of the effects of replacing two surface charges by all the other amino acids resulted in the identification of six variants that are more stable than the wild-type enyme in 80% DMF. Substitution of Arg 45 and Arg 78 by large hydrophobic residues (Phe, Tyr, Leu, Ile) stabilizes the protein in the presence of the organic solvent. The effects of substitutions were found to be additive; substitution of both charged residues by neutral ones yields a variant that is 10 times more stable than the wild-type enzyme in 80% DMF (Martinez, P. and Arnold, F. H., in preparation).

The catalytic activity of an enzyme in organic solvents can also be enhanced by protein engineering. A subtilisin E variant with two mutations, Asn 218 → Ser and Glu 103 → Arg (identified by screening randomly mutagenized colonies for improved protease activity in the presence of DMF), is ten times more active than the wild-type enzyme in 20% DMF [103]. The effects of the individual mutations on the free energies of transition state stabilization were found to be additive in both aqueous solution and the mixed solvent.

CONCLUSIONS

Enzymes have been engineered to improve thermostability with a considerable degree of success; substantial stability enhancements can be achieved through seemingly minor alterations in amino acid sequence. By engineering the amino acid sequence to compensate for interactions in a new solvent environment, it should be possible to obtain comparable improvements in enzyme stability in nonaqueous solvents. Thus the future for enzyme engineering to create catalysts that will function in polar nonaqueous solvents appears promising.

The mechanisms of stabilization discussed in this chapter and the design rules presented in Table 2 are proposals meant to be tested by future work. There are simply not enough data yet on engineering stability in nonaqueous solvents to permit us to draw conclusions regarding the most effective approach to enzyme design. Those data will be forthcoming as more researchers document the effects of particular substitutions on enzyme stability in solvents other than water. We emphasize that the design rules of Table 2 are guidelines to the types of substitutions that should enhance stability, not hard and fast rules that can be expected to apply in all cases. It is notoriously difficult to make an amino acid substitution that affects only a single type of interaction. Furthermore, the relative importance of the various interactions will change from protein and even from domain to domain in a single protein.

One might wonder to what extent an enzyme would have to be modified to optimize its stability in polar nonaqueous solvents, an environment that may be significantly different from the natural aqueous one. If a large number of substitutions are beneficial, a second question arises: How many substitutions will be tolerated before the enzyme will no longer fold or be produced in the expression system used? The answers are, of course, unknown, but in this regard the results of Wong, Pantoliano and their coworkers with subtilisin BPN′ are promising. In this case, six substitutions were made, resulting in a large improvement in stability and no inhibition of the enzyme's catalytic activity. A comparison of homologous enzyme sequences indicates that large numbers of amino acid substitutions, even 30% of the sequence, can be tolerated in very similar folded structures.

The application of protein engineering to create enzyme catalysts capable of functioning in nonnatural environments has been successfully demonstrated with the enzyme subtilisin, and more such demonstrations are likely to come in the near future. As the stabilization mechanisms discussed in this chapter are verified and extended through the study of mutant enzymes, we hope to develop a rational basis for designing enzymes with improved stability in nonaqueous solvents. With continued success, the technological applications of nonaqueous solvent-stable enzymes will multiply.

REFERENCES

1. Ahern, T. J., and Klibanov, A. M. *Science, 228*, 1280-1284 (1985).
2. Klibanov, A. M. *Chemtech*, June, 354-359 (1986).
3. Dordick, J. S. *Enzyme Microb. Technol. 11*, 194-211 (1989).
4. Laane, C., Boeren, S., Vos, K., and Veeger, C. *Biotech. Bioeng. 30*, 81-87 (1987).
5. Halling, P. J. *Biotechnol. Adv. 5*, 47-84 (1987).
6. Levinthal, C. *Proteins: Struct. Funct. Gen. 1*, 2-3 (1986).
7. Zaks, A., and Klibanov, A. M. *J. Biol. Chem. 263*, 8017-8021 (1988).
8. Laane, C., Boeren, S., and Vos, K. *Trends Biotechnol. 3*, 251-252 (1985).
9. Reslow, M., Aldercreutz, P., and Mattiason, B. *Appl. Microbiol. Biotech. 26*, 1-8 (1987).
10. Nosoh, Y., and Sekiguchi, T. *Biocatalysis, 1*, 257-273 (1988).
11. Menendez-Arias, L., and Argos, P. *J. Mol. Biol. 206*, 397-406 (1989).
12. Pantoliano, M. W., Ladner, R. C., Bryan, P. N., Rollence, M., L., Wood, J. F., and Poulos, T. L. *Biochemistry, 26*, 2077-2082 (1987).
13. Yutani, K., Ogasahara, K., Tsujita, T., and Sugino, Y. *Proc. Natl. Acad. Sci. USA, 84*, 4441-4444.
14. Imanaka, T., Shibazaki, M., and Takagi, M. *Nature, 324*, 695-697 (1986).
15. Matsumura, M., Becktel, W. J., Levitt, M., and Matthews, B. W. *Proc. Natl. Acad. Sci. USA, 86*, 6562-6566 (1989).
16. Matsumura, M., Signor, G., and Matthews, B. W. *Nature, 342*, 291-293 (1989).
17. Gavish, B., Gratton, E., and Hardy, C. J. *Proc. Natl. Acad. Sci. USA, 80*, 750-754 (1983).
18. Rupley, J. A., Gratton, E., and Careri, G. *Trends Biochem. Sci. 8*, 18-22 (1983).
19. Finney, J. L., and Poole, P. L. *Comments Mol. Cell. Biophys. 2*, 129-151 (1984).
20. Parak, F. *Methods Enzymol. 127*, 196-206 (1986).
21. Jansen, H., Matuszak, E., Goldammer, E. V., and Wenzel, H. R. *Z. Naturforsch C, 43c*, 285-293 (1988).
22. Careri, G., Giansanti, A., and Gratton, E., *Biopolymers, 18*, 1187-1203 (1979).
23. Poole, P. L., and Finney, J. L. *Int. J. Biol. Macromol. 5*, 308-310 (1983).

24. Poole, P. L., and Finney, J. L. *Biopolymers, 22*, 255-260 (1983).
25. Picullel, L., and Halle, B. *J. Chem. Soc. Faraday Trans.* 1, *82*(1), 401-414, (1986).
26. Rupley, J. A., Yang, P. -H., and Tollin, G. In *Water in Polymers* (S. P. Rowland ed.), American Chemical Society, Washington, DC 1980, pp. 111-132.
27. Yang, P. H., and Rupley, J. A. *Biochemistry, 18*, 2654-2661 (1979).
28. Careri, G., Gratton E., Yang, P.-H., and Rupley, J. A. *Nature, 284*, 572-573, (1980).
29. Finney J. L., Goodfellow, J. M., and Poole, P. L. In *Structural Molecular Biology*, (D. B., Davies, S. Danyluk and W. Saenger, eds.), Plenum Press, New York, 1982, pp. 387-426.
30. Gol'dovskii, A. M., and Smolyak, J. V. *Zh. Evol. Biokh. Fiziol. 19*, 300-302 (1983).
31. Poole, P. L., Barlow, D. J., Wolanczyk, J. P., and Finney J. L. *Proceedings of the 1st European Conference on the Spectroscopy of Biological Molecules*, John Wiley & Sons, New York, 1985, pp. 208-212.
32. Bone, S., and Pethig, R. *J. Mol. Biol. 157*, 571-575 (1982).
33. Marchetti, P. S., Bhattacharyya, L., Ellis, P. D., and Brewer, C. F. *J. Magn. Resonance, 80*, 417-426 (1988).
34. Woodward, C., Simon, I., and Tuchsen, E. *Mol Cell. Biochem. 48*, 135-160 (1982).
35. Englander, S. W., and Kallenbach, N. R. *Q. Rev. Biophys. 16*, 521-655 (1984).
36. Englander, J. J., Englander, S. W., Louie, G., Roder, H., Tran, T., and Wand, A. J. In *Structure and Expression*, Vol. 1; *From Proteins to Ribosomes* (R. H. Sarma and M. H. Sarma, eds.), Adenine Press, Schenectady, New York, 1988, pp. 107-117.
37. Karplus, M., and McCammon, J. A. *Annu. Rev. Biochem. 53*, 263-300 (1983).
38. McCammon, J. A., and Harvey, S. C. *Dynamics of Proteins and Nucleic Acids*, Cambridge University Press, New York, 1987.
39. Swaminathan, S., Ichiye, T., van Gunsteren, W. F., and Karplus, M. *Biochemistry, 21*, 5230-5241 (1983).
40. MacKerell, A. D. Jr., Nilsson, L., Rigler, R., and Saenger, W. *Biochemistry, 27*, 4547-4556 (1988).
41. Tanford, C. *Adv. Protein Chem. 24*, 1-95 (1970).
42. Edsall, J. T., and McKenzie, H. A. *Adv. Biophys. 16*, 53-183 (1983).
43. Wolfenden, R., Andersson, L., Cullis, P. M., and Southgate, C. C. B. *Biochemistry, 20*, 849-855 (1981).
44. Radzicka, A., and Wolfenden, R. *Biochemistry, 27*, 1664-1670 (1988).
45. Tanford, C. *J. Am. Chem. Soc. 84*, 4240-4247 (1962).
46. Nozaki, Y., and Tanford, C. *J. Biol. Chem. 246*, 2211-2217 (1971).
47. Bull, H. B., and Breese, K. *Arch. Biochem. Biophys. 161*, 665-670 (1974).
48. Fendler, J., Nome, F., and Nagyvary, J. *J. Mol. Evol. 6*, 215-232 (1975).
49. Matsumura, M., Becktel, W. J., and Matthews, B. W. *Nature, 334*, 406-410 (1988).
50. Hvidt, A. *J. Theor. Biol. 50*, 245-252 (1975).

51. Kellis, J. T., Jr., Nyberg, K., Sali, D., Fersht, A. R. *Nature, 333*, 784-786 (1988).
52. Kellis, J. T., Jr., Nyberg, K., and Fersht, A. R. *Biochemistry, 28,* 4914-4922 (1989).
53. Karpusas, M., Baase, W. A., Matsumura, M., and Matthews, B. W. *Proc. Natl. Acad. Sci. USA, 86,* 8237-8241 (1989).
54. Pantoliano, M. W., Whitlow, M., Wood, J. F., Dodd, S. W., Hardman, K. D., Rollence, M. L., and Bryan, P. N. *Biochemistry, 28,* 2705-7213 (1989).
55. Privalov, P. L. and Gill, S. J. *Adv. Protein Chem. 39,* 191- 234 (1988).
56. Sandberg, W. S., and Terwillinger, T.C. *Science, 245,* 54-57 (1989).
57. Damodaran, S., and Song, K. B. *J. Biol. Chem. 261,* 7220- 7222 (1986).
58. Chothia, C. *J. Mol. Biol. 105,* 1-14 (1976).
59. Rees, D. C., DeAntonio, L., and Eisenberg, D. *Science, 245,* 510-513. (1989).
60. Finney, J. L., Gellatly, B. J., Golton, I. C., and Goodfellow, J. *Biophys. J. 32,* 17-32 (1980).
61. Alber, T., Dao-pin, S., Wilson, K., Wozniak, J. A., Cook, S. P., and Matthews, B. W. *Nature, 330,* 41-46 (1987).
62. Grutter, M. G., Gray, T. M., Weaver, L. H., Alber, T., Wilson, K., and Matthews, B. W. *J. Mol. Biol. 197,* 315-329. (1987).
63. Fersht, A. R. *Trends Biochem. Sci. 12,* 301-304 (1987).
64. Fersht, A. R., Shi, J. -P., Knill-Jones, J., Lowe, D. M., Wilkinson, A. J., Blow, D. M., Brick, P., Carter, P., Waye, M. M. Y., and Winter, G. *Nature, 314,* 235-238 (1985).
65. Bryan, P. N., Rollence, M. L., Pantoliano, M. W., Wood, J., Finzel, B. C., Gilliland, G. L., Howard, A. J., and Poulos, T. L. *Proteins: Struct. Funct. Gen. 1,* 326-334 (1986).
66. Klotz, I. M., and Franzen, J. S. *J. Am. Chem. Soc. 84,* 3461-3466 (1962).
67. Sneddon, S. F., Tobias, D. J., and Brooks, C. L. III. *J. Mol. Biol. 209,* 817-820 (1989).
68. Marqusee, S., and Baldwin, R. L. *Proc. Natl. Acad. Sci., USA, 84,* 8898-8902 (1987).
69. Barlow, D. J., and Thornton, J. M. *J. Mol. Biol. 168,* 867-885 (1983).
70. Hwang, J.-K, and Warshel, A. *Nature, 334,* 270-272 (1988).
71. Warshel, A., Russell, S. T., and Churg, A. K. *Proc. Natl. Acad. Sci., USA, 81,* 4785-4789 (1984).
72. Nicholson, H., Becktel, W. J., and Matthews, B. W. *Nature, 336,* 651-656 (1988).
73. Voordouw, G., Milo, C., and Roche, R. S. *Biochemistry, 15,* 3716-3724 (1976).
74. Pantoliano, M. W., Whitlow, M., Wood, J. F., Rollence, M. L., Finzel, B. C., Gilliland, G. L., Poulos, T. L., and Bryan, P. N. *Biochemistry, 27,* 8311-8317 (1988).
75. Gilson, M. K., and Honig, B. *Proteins: Struct. Funct. Gen. 4,* 7-18 (1988).
76. Rashin, A. A., and Honig, B. H. *J. Mol. Biol. 173,* 515-521 (1984).
77. Villafranca, J. E., Howell, E. E., Voet, D. H., Strobel, M. S., Ogden, R. C., Abelson, J. N., and Kraut, J. *Science, 222,* 782-788 (1983).
78. Pabo, C. O., and Sukchanek, E. G. *Biochemistry, 25,* 5987-5991 (1986).

79. Sauer, R. T., Hehir, K., Stearman, R. S., Weiss, M. A., Jeitler-Nilsson, A., Suchanek E. G., and Pabo, C. O. *Biochemistry 25*, 5992-5998 (1986).
80. Mitchinson, C., and Wells, J. A. *Biochemistry, 28*, 4807-4815 (1989).
81. Wetzel, R. *Trends Biochem. Sci. 12*, 478-482 (1987).
82. Pjura, P. E., Matsumura, M., Wozniak, J. A., and Matthews, B. W. *Biochemistry, 29*, 2592-2598 (1990).
83. Matthews, B. W., Nicholson, H., and Becktel, W. J. *Proc. Natl. Acad. Sci. USA, 84*, 6663-6667 (1987).
84. Hecht, M. H., Sturtevant, J. M., and Sauer, R. T. *Proteins: Struct. Funct. Gen. 1*, 43-46 (1986).
85. Wetzel, R., Perry, L. J., Baase, W. A., and Becktel, W. J. *Proc. Natl. Acad. Sci. USA, 85*, 401-405 (1988).
86. Matsumura, M., Yasumura, S., and Aiba, S. *Nature, 323*, 356-358 (1986).
87. Zaks, A., and Klibanov, A. M. *J. Biol. Chem. 263*, 3194-3201 (1988).
88. Clark, D. S, Creagh, L., Skerker, P., Guinn, M., Prausnitz, J., and Blanch, H. W. *ACS Symp. Ser. 392*, 104-114 (1989).
89. Burke, P. A., Smith, S. P., Bachovchin, W. W., and Klibanov, A. M. *J. Am. Chem. Soc. 111*, 8290-8291 (1989).
90. Vihinen, M., *Protein Eng. 1*, 477-480 (1987).
91. Klibanov, A. M. *Trends Biochem. Sci. 14*, 141-144 (1989).
92. Russell, A. J., Thomas, P. G., and Fersht, A. R. *J. Mol. Biol. 193*, 803-813 (1987).
93. DeMarco, A., Lecomte, J. T. J., and Llinas, M. *Eur. J. Biochem. 119*, 483-490 (1981).
94. Arnold, F. H. *Ann. N. Y. Acad. Sci. 542*, 30-36 (1989).
95. Teeter, M. M., Mazer, J. A., and L'Italien, J. J. *Biochemistry, 20*, 5437-5443 (1981).
96. Lecomte, J. T. J., Kaplan, D., Llinas, M., Thunberg, E., and Samuelsson, G. *Biochemistry, 26*, 1187-1193 (1987).
97. Clore, G. M., Sukumaran, D. K., Gronenborn, A. M., Teeter, M. M., Whitlow, M., and Jones, B. L. *J. Mol. Biol. 193*, 571-578 (1987).
98. Arnold, F. H. *Protein Eng. 2*, 21-25 (1988).
99. Hendrickson, W. A., and Teeter, M. M. *Nature, 290*, 107-112 (1981).
100. Barlow, D. J., and Thornton, J. M. *Biopolymers, 25*, 1717- 1733 (1986).
101. Teeter, M. M. *Proc. Natl. Acad. Sci. USA, 81*, 6014-6018 (1984).
102. Wong, C. -H., Chen, S. -T., Hennen, W. J., Bibbs J. A., Wang, Y. F., Liu, J. L. -C., Pantoliano, M. W., Whitlow, M., and Bryan, P. N. *J. Am. Chem. Soc. 112*, 945-953 (1990).
103. Chen, K., Robinson, A. C., Van Dam, M., Martinez, P., Economou, C., and Arnold, F. H. *Biotechnol Progr.*, in press.

3

Applications of Enzymatic Aldol Reactions in Organic Synthesis

Mark David Bednarski

Department of Chemistry, University of California at Berkeley, Berkeley, California

INTRODUCTION

The development of methods for the stereoselective formation of carbon-carbon bonds using the aldol reaction is a current topic of interest in organic synthesis [1-6]. Many successful strategies rely on chiral auxiliaries [7-15], and a few examples exist that use organometallic [16-19] catalysts.

The use of enzymes in synthetic organic chemistry is only now being explored [20-41]. This chapter discusses the utility of readily available carbon-carbon bond forming enzymes as catalysts for the asymmetric aldol reaction.

ENZYMES THAT FORM CARBON–CARBON BONDS

Three general types of enzymes have been applied for the formation of carbon-carbon bonds in organic synthesis: aldolases, synthetases, and transketolases. Aldolases are a class of enzymes that catalyze the stereoselective construction and degradation of carbon-carbon bonds in monosaccharides [42-46]. Synthetases catalyze an irreversible aldol reaction of activated enols such as phosphoenolpyruvate (PEP) with aldoses to form complex monosaccharides [47]; transketolases catalyze the transfer of a hydroxyketo group of a keto sugar to an aldose [48]. These enzymes are discussed individually below, with an outline of their use in organic synthesis.

d-FRUCTOSE-1,6,-DIPHOSPHATE ALDOLASE (FDP)

d-Fructose-1,6-diphosphate (FDP) aldolase from rabbit muscle (E. C. 4.1.2.13) catalyzes the equilibrium condensation of dihydroxyacetone phosphate (1) (DHAP) with d-glyceraldehyde-3-phosphate (2) (G-3-P) to form d-fructose-1, 6-biphosphate (3) (FDP) (Scheme 1) [42–44]. The equilibrium

constant for this reaction is $K = 10^4$ M–1 in favor of the formation of FDP. The stereoselectivity of the reaction is absolute; the configuration of the vicinal diols at C-3′ and C-4 is always threo (i.e. d-glycero). Although there is a significant discrimination (20:1) between the antipodes of the natural substrate (i.e., d-and l-G-3-P) this selectivity extends only a few unnatural substrates [33].

Substrate Specificity

FDP aldolase accepts a wide range of aldehydes in place of its natural substrate G-3-P [33], thus permitting the synthesis of carbohydrates such as nitrogen-containing sugars [34,49], deoxy sugars [50,51], fluorosugars [51], and C-8, C-9 sugars [52]. More than 75 aldehydes have been identified as substrates based on enzymatic assay, and many of the aldol adducts have been isolated and characterized [53-77]. In general, unhindered aliphatic, α-heteroatom-substituted, and differentially protected alkoxy aldehydes are substrates; severely hindered aliphatic aldehydes such as pivaldehyde do not react with FDP aldolase, nor do α, β-unsaturated aldehydes, although these compounds do not inhibit the enzyme. Aromatic aldehydes are either poor substrates or unreactive. Some examples are summarized in Table 3.1.

The requirement for the electrophilic component (DHAP) is much more stringent than for the nucleophilic component; so far investigators have dem-

Table 1 Products Synthesized by Reacting the Substrate with Dihydroxyacetone Phosphate in the Presence of FDP Aldolase

Substrate	Product	Reference
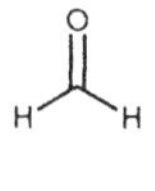	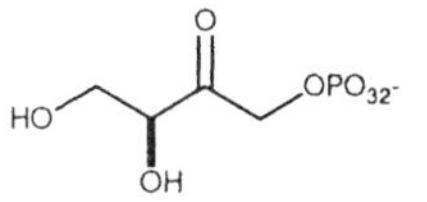	33
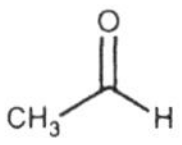	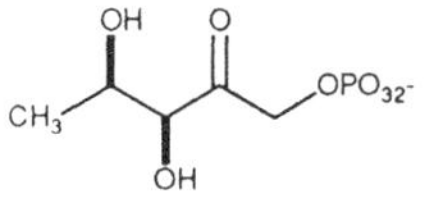	60,69
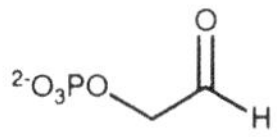	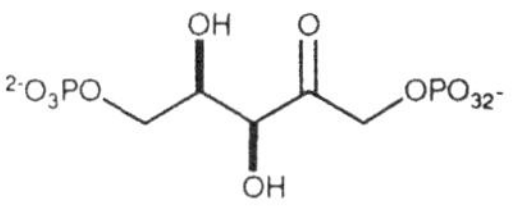	59,70
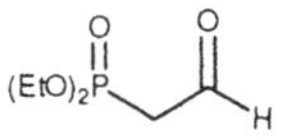	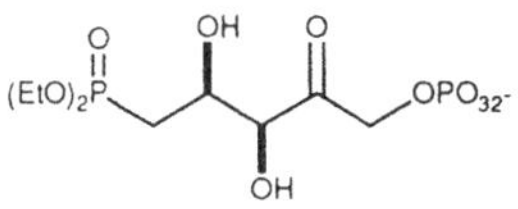	66
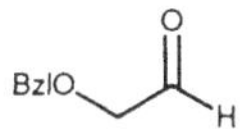	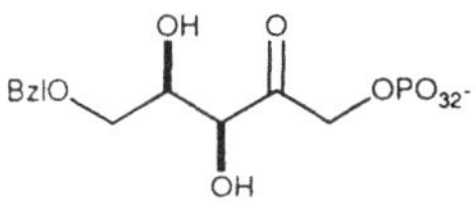	33
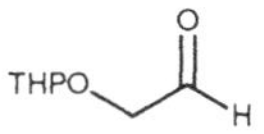	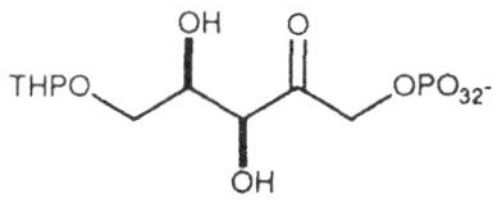	33
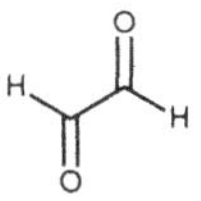	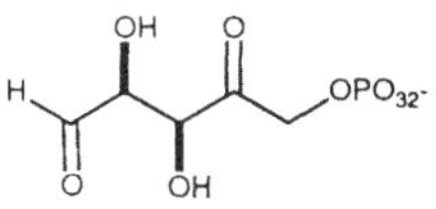	66
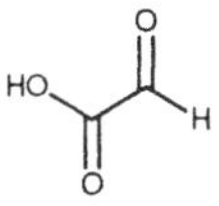	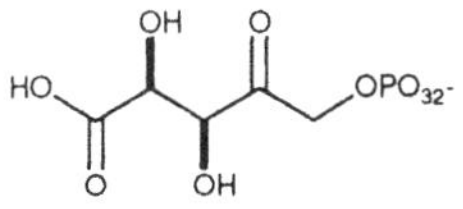	66
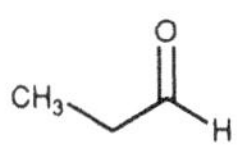	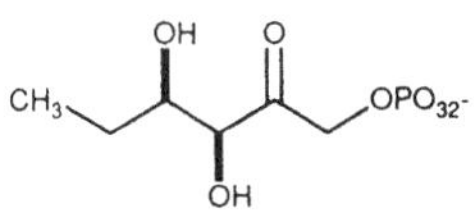	33,61,72

Table 1 Continued

Substrate	Product	Reference
D- Threose	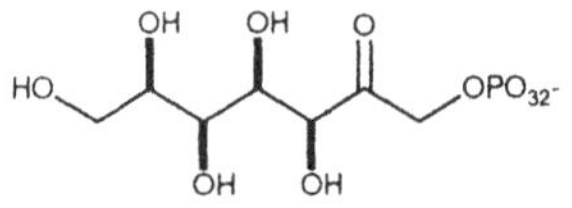	60
L- Threose	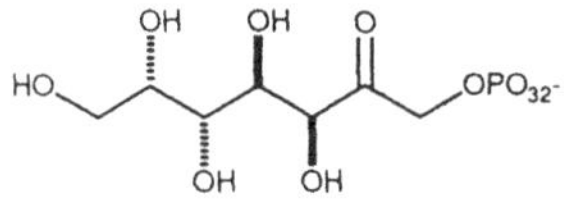	55
D-Erythrose	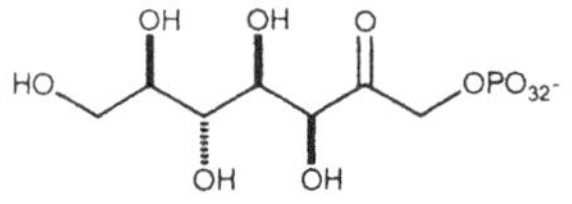	33,59,62,65
D-Erythrose-4-P	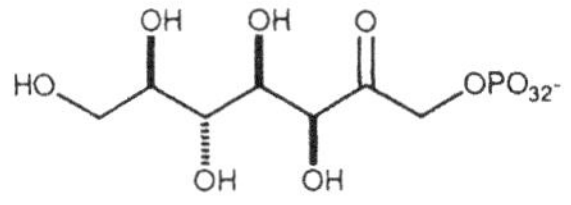	71,73
L-Erythrose	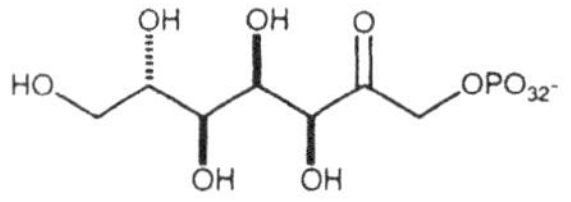	62
D-Ribose	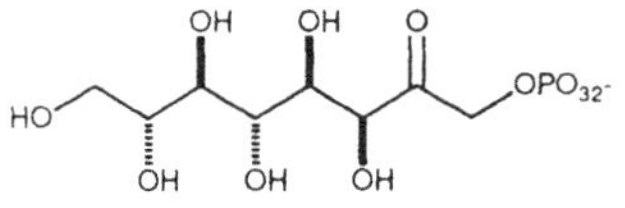	53,57
D-Riboses-5-P	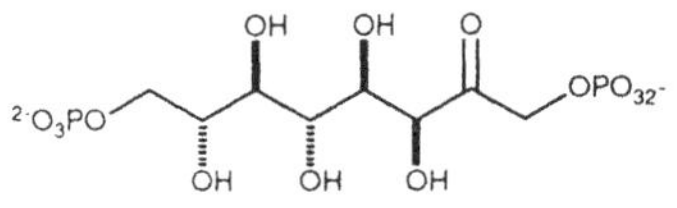	52,58,59,63
2-deoxy-D-Ribose-5-P	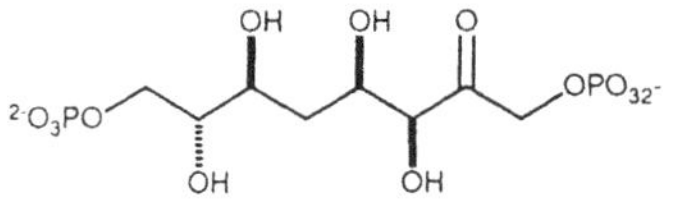	52
L-Arabinose	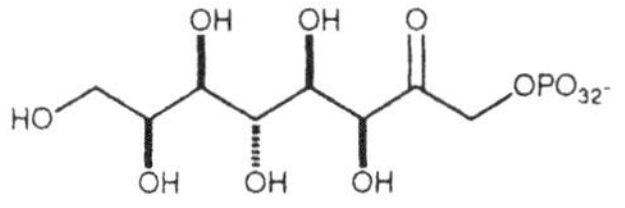	53

Table 1   Continued

Substrate	Product	Reference
D-Arabinose-5-P	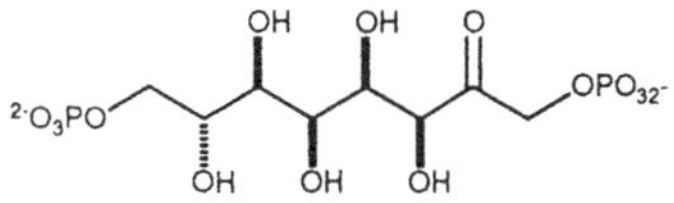	52,58,63
D-Lyxose	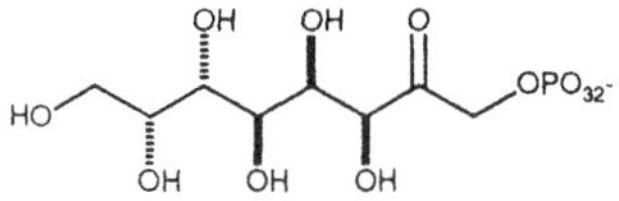	53
D-Xylose	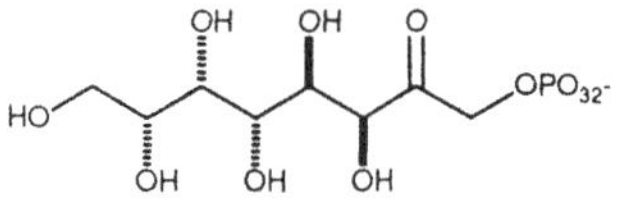	53
D-Glucose-6-P	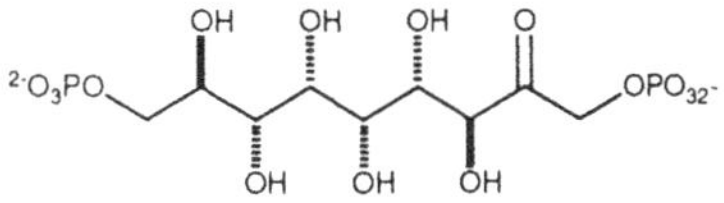	52
2-deoxy-D-Glucose-6-P	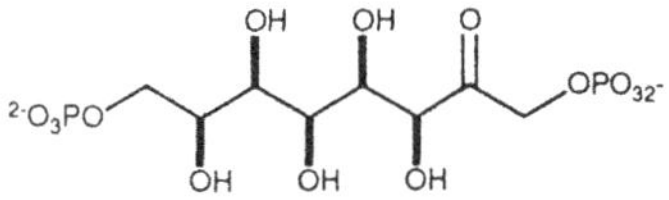	52
D-Glucosamine-6-P	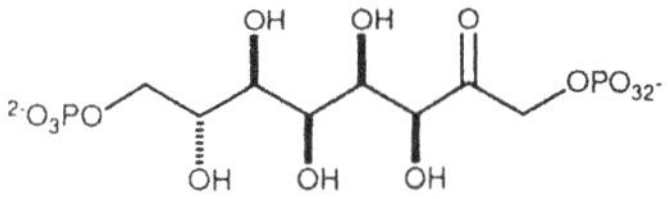	52
D-Galactose-6-P	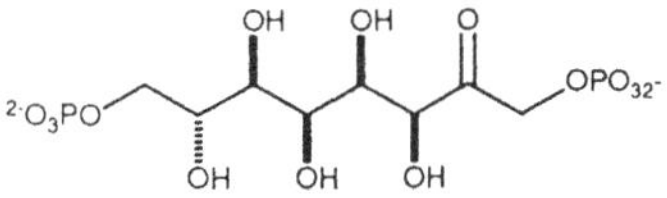	52
D-Mannose-6-P	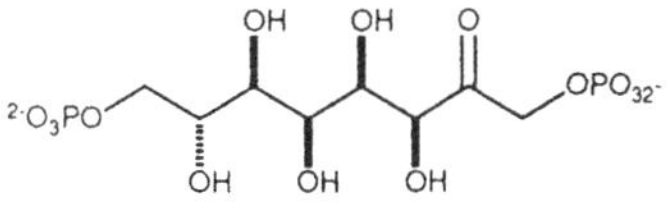	52

Table 1 Continued

Substrate	Product	Reference
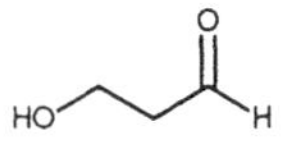	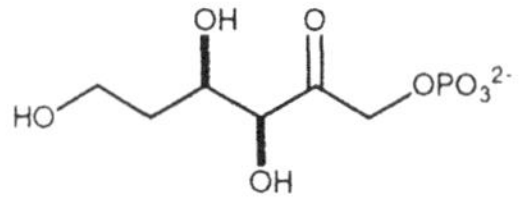	33
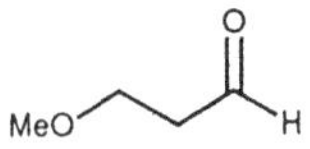	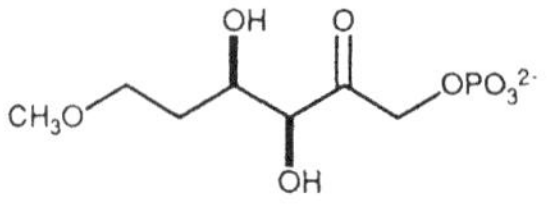	72
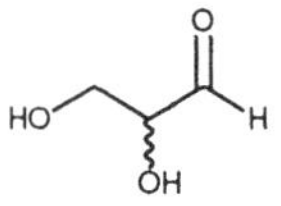	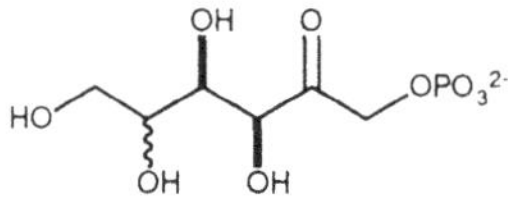	50
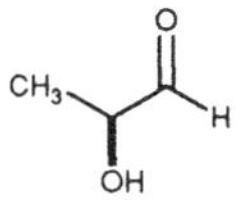	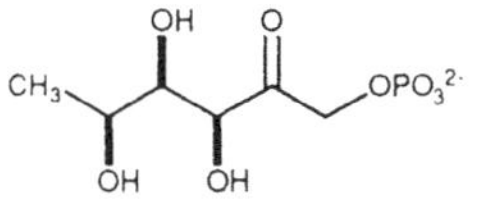	33
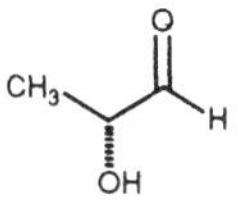	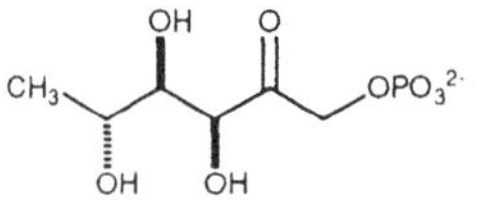	33
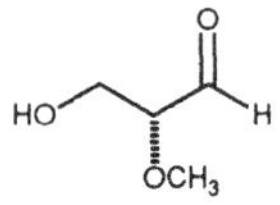	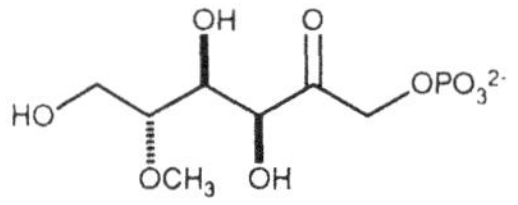	33
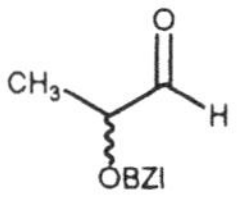	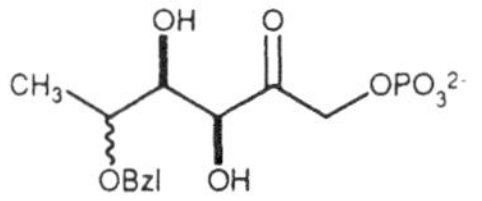	33
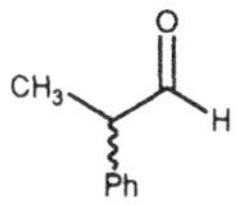	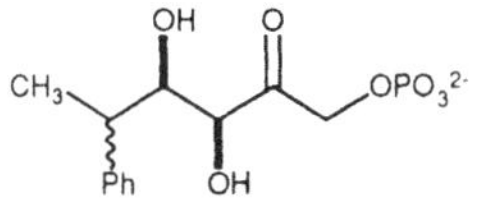	33
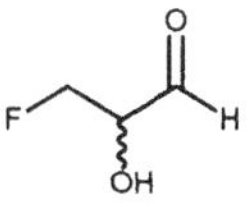	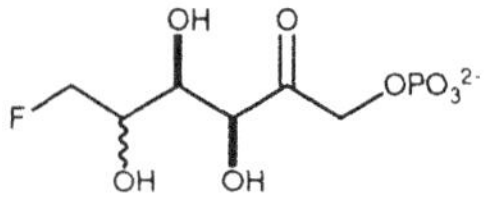	51

Table 1 Continued

Substrate	Product	Reference
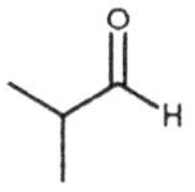	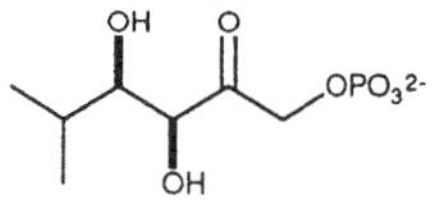	66
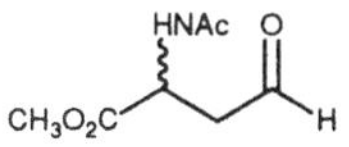	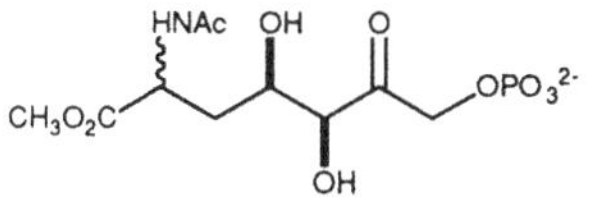	66
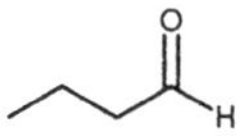	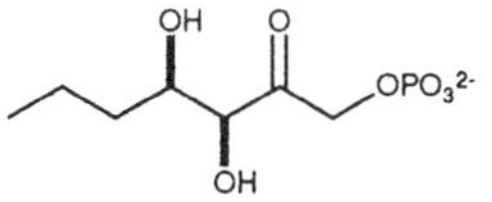	66,33
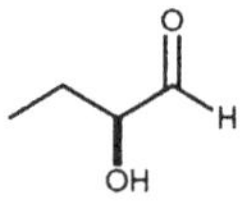	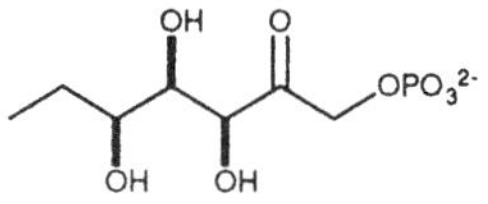	51,49,68,69
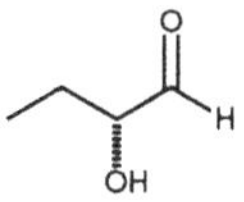	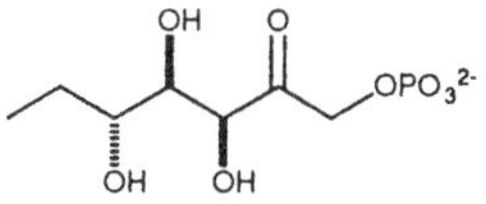	49,51,68,69
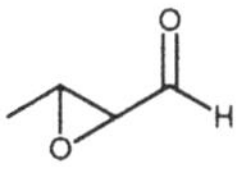	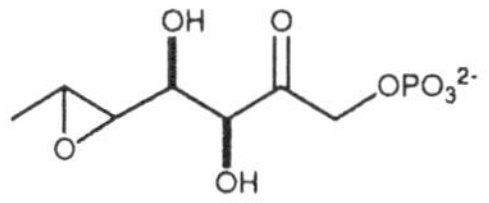	66
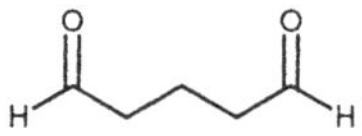	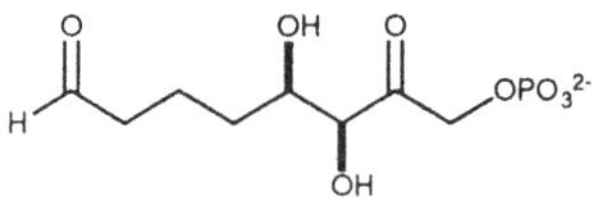	66
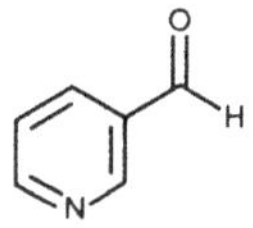	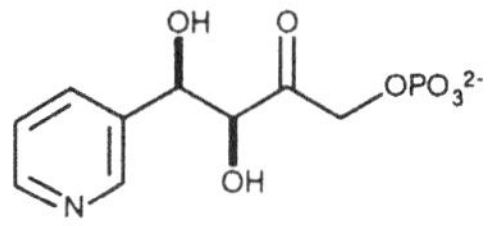	33,66
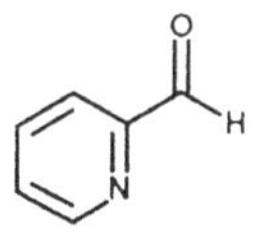	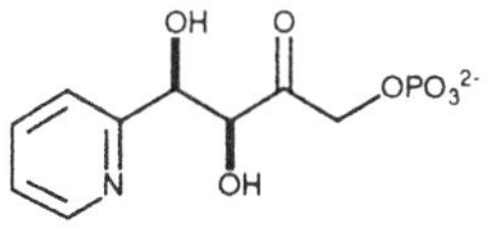	33,66
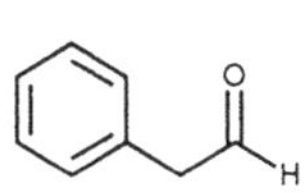	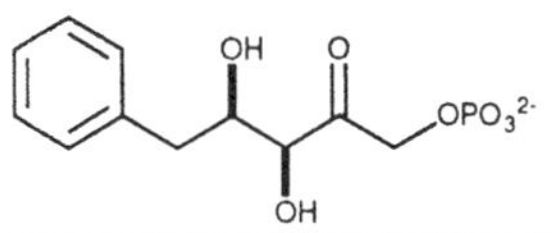	33,66

onstrated that only 1,3-dihydroxy-2-butanone-3-phosphate and 1,4-dihy-droxy-3-butanone-1-phosphonate are substrates (Table 3.2) [33].

Preparation of Dihydroxyacetone Phosphate (DHAP).

DHAP is an essential substrate of aldolase-catalyzed reactions and a simple preparation of this compound is essential for developing the use of this enzyme in asymmetric synthesis. DHAP may be generated by three procedures: (1) in situ from fructose-1,6-diphosphate with the enzymes, aldolase, and triosephosphate isomerase (EC 5.3.1.1.), [33]; (2) from the dimer of di-hydroxyacetone by chemical phosphorylation with $POCl_3$, [33,66]; and (3) from dihydroxyacetone by enzymatic phosphorylation using PEP and glycerol kinase (EC 2.7.1.30) [33,50,76]. The in situ generation of DHAP from FDP is the most convenient method. This reaction does not however, go to comple-tion, and the presence or excess FDP can complicate the isolation of products that are negatively charged. In these cases the chemical synthesis of DHAP is the method of choice.

A mixture of dihydroxyacetone and inorganic arsenate may replace DHAP [49], and this mixture has been used in synthesis by Wong and coworkers. A dihydroxyacetone arsenate monoester probably forms in the rate-determining step, and is consumed in a fast, irreversible aldol reaction. The irreversibility imposed by this method may be useful with slowly reacting substrates, but the toxicity of arsenate limits its usefulness. Vanadate does not operate as a phos-phate mimic in FDP aldolase catalyzed reactions [49].

The phosphate group (derived from DHAP) of the aldol adducts facilitates the purification of aldol products by ion-exchange chromatography or by pre-cipitation as their barium or silver salts. Enzymatic methods using acid or al-kaline phosphatase or chemical methods using acid or base also allow cleavage of this group [33].

Enzyme Characteristics

Several characteristics of FDP aldolase make it a useful enzyme for use in syn-thesis. Commercial preparations of the enzyme are inexpensive (it costs 4 cents to produce one millimole of product per minute) and it has a reasonable specific activity (60 U/mg of protein). FDP aldolase requires no metal ions or cofactors, it is stable in the presence of oxygen and added organic cosolvents, and it is not inhibited to a significant extent by the natural product (i.e., FDP). The enzyme has been used in soluble or immobilized forms, or enclosed within a dialysis membrane [33].

Table 2 Relative Rates of DHAP Analogues in Reactions Catalyzed by FDP Aldolase

R_1	R_2	Vrel
$HOCH_2$	$CH_2OPO_3^{2-}$	100
$HOCH_2$	$CHMeOPO_3^{2-}$	10
$HOCH_2$	$CH_2CH_2PO_3^{2-}$	10
$HOCH_2$	CH_2SO_3H	<0.1
$HOCH_2$	CH_2OH	<0.1
Me	$CH_2OPO_3^{2-}$	<0.1
N_3CH_2	$CH_2OPO_3^{2-}$	<0.1
$AcNHCH_2$	$CH_2OPO_3^{2-}$	<0.1
$HOCH_2$	$CH_2OPO_3^{2-}$	<0.1
$ClCH_2$	$CH_2OPO_3^{2-}$	0
$BrCH_2$	$CH_2OPO_3^{2-}$	0

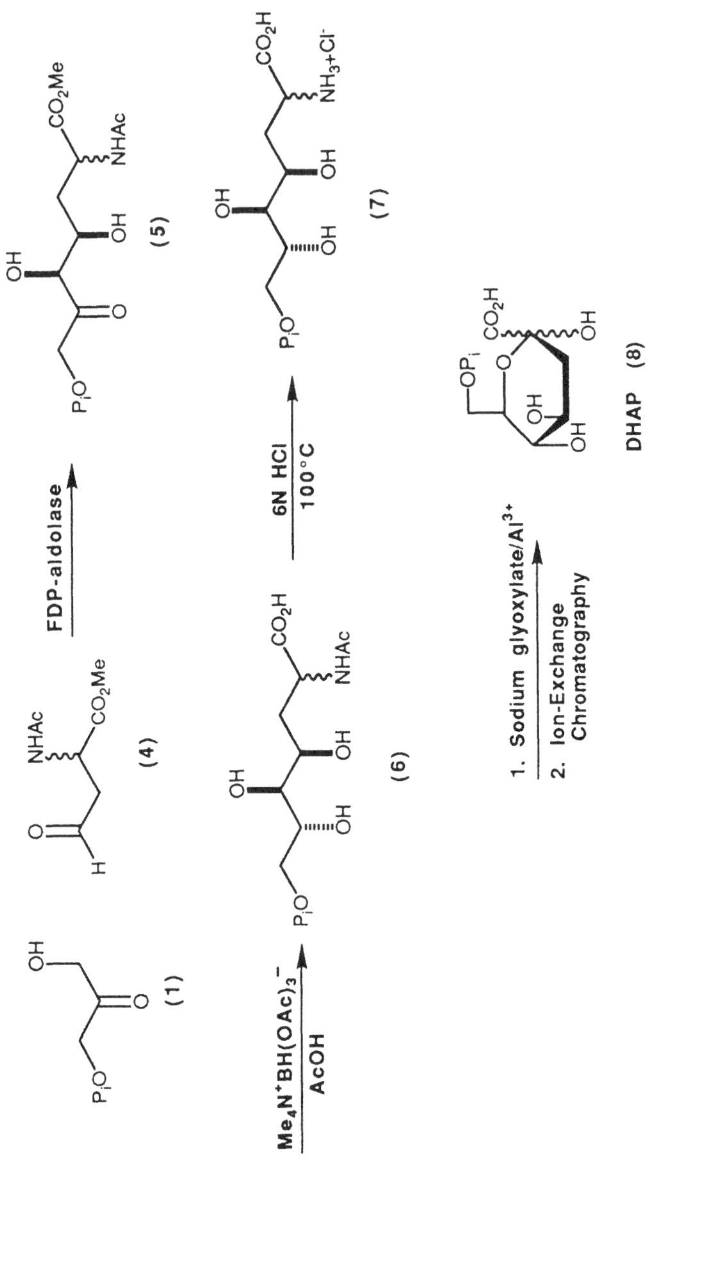
(1)
(4)
FDP-aldolase
(5)
Me$_4$N$^+$BH(OAc)$_3^-$
AcOH
(6)
6N HCl
100°C
(7)
1. Sodium glyoxylate/Al^{3+}
2. Ion-Exchange Chromatography
DHAP (8)

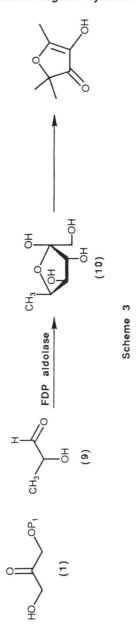

Scheme 3

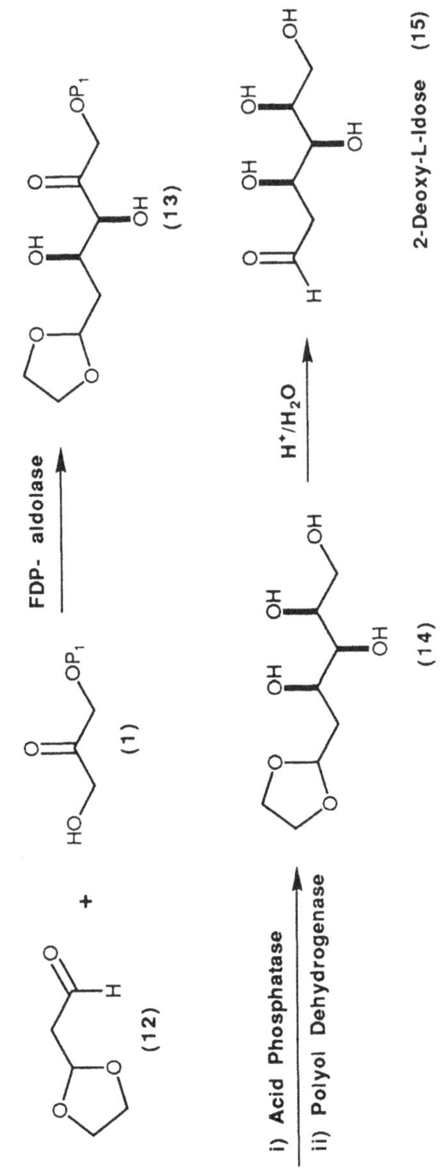
(12)
(1)
FDP- aldolase
(13)
i) Acid Phosphatase
ii) Polyol Dehydrogenase
(14)
H+/H2O
2-Deoxy-L-Idose (15)

Examples of FDP Aldolase in Organic Synthesis

The biologically active monosaccharide 3-deoxy-d-*arabino*-heptulosonic acid 7-phosphate (8) (DAHP) is an important intermediate in the biosynthesis of aromatic amino acids in plants (the shikimate pathway). As shown in Scheme 2 (see page 92), this compound has been produced in a combined chemical and enzymatic synthesis from racemic *N*-acetylaspartate β-semialdehyde (4) and dihydroxyacetone phosphate (1) [67]. The four-step synthesis proceeds in an overall yield of 13% (37% for the aldolase reaction). The enzymatic step generates the required, enantiomerically pure, syn aldol adduct compound having threo stereochemistry (5). In view of the broad range of substrates tolerated by FDP aldolase, this method may be applicable to the production of analogues of DAHP.

Other monosaccharide derivatives have also been synthesized using FDP aldolase as a catalyst. Deoxyfructose derivatives such as a compound (10) were, for example, synthesized using FDP aldolase catalyzed addition of DHAP to lactaldehyde (9). The aldol adduct (10) was chemically converted to furaneol (11), an important flavoring agent (Scheme 3; see page 97) [50]. Other deoxy sugars such as 2-deoxy-l-idose (15) are also available using FDP aldolase. The condensation of aldehyde 12 with DHAP gives 13, which upon

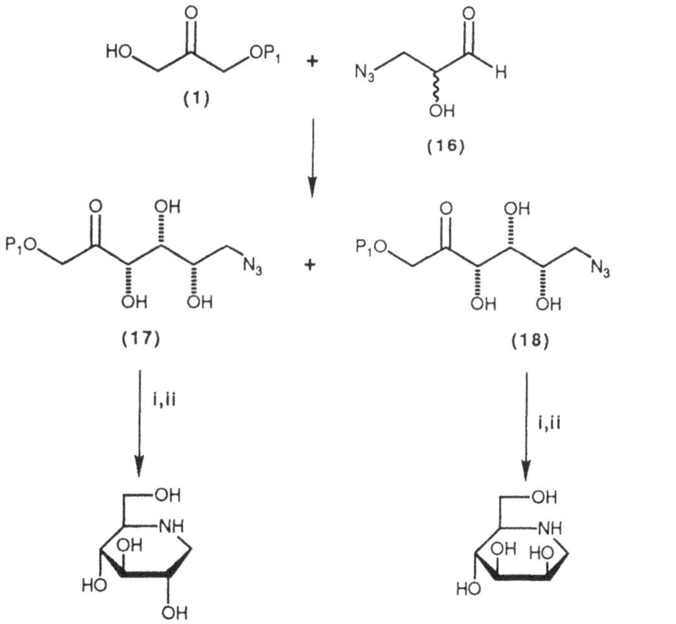

Deoxynojirimycin   (19)   Deoxymannonjirimycin   (20)

i) Phosphatase (or H⁺)   ii) H₂/Pd

dephosphorylation reduction, and deprotection yields the deoxyidose 15 (Scheme 4; see page 93) [50,51]. Two groups have also used aldolase to synthesize alkaloids [34,78]. Aldehyde 16 was condensed with DHAP (1) using aldolase as a catalyst to give 17 and 18, respectively (Scheme 5; see page 95). These compounds were chemically converted into deoxynojirimycin (19) and deoxymannojimycin (20) as shown [79].

N-ACETYLNEURAMINIC ACID ALDOLASE (E.C.4.1.3.3)

N-Acetylneuraminic acid aldolase catalyzes the reversible aldol condensation of pyruvate (23) and *N*-acetylmannosamine (22) (ManNAc) to form *N*-acetyl-neuraminic acid (24) (NeuAc, *N*-acetyl-5-amino-3,5-dideoxy-d-glycero-galacto-2-nonulopyronic acid) (Scheme 6) [80–83]. In vivo the enzyme has a

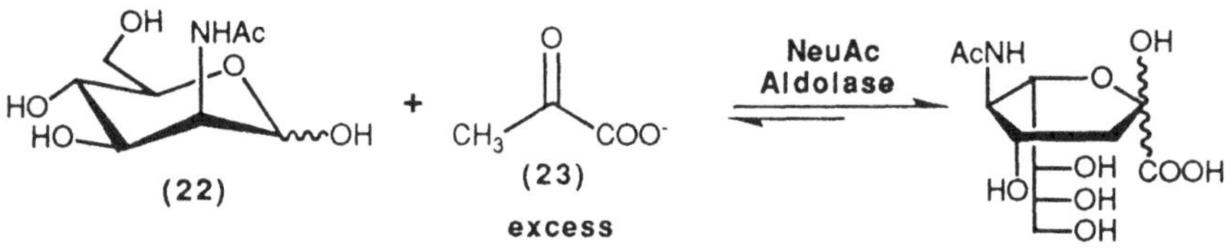

catabolic function, and the equilibrium for this reaction is near unity; the presence of excess pyruvate can shift this equilibrium. NeuAc and other derivatives of neuraminic acid are termed sialic acids. These compounds are found at the termini of mammalian glycoconjugates and play an important role in cellular recognition [84,89]. The production of analogues of NeuAc is a point of great interest to synthetic and medicinal chemists. The enzymatic approach has not been fully explored, but it may be a practical alternative to the chemical synthesis of certain sialic acids [89].

Reactions catalyzed by NeuAc aldolase have produced several grams of NeuAc after purification by ion-exchange chromatography [90–92]. Although the specificity for pyruvate appears to be absolute, the results of enzymatic assays suggest that NeuAc aldolase accepts a range of substrates in place of ManNAc [80,81,90–97]. Several groups have taken advantage of this observation to synthesize and isolate derivatives of NeuAc. General observations from these syntheses and from assay results suggest that NeuAc aldolase will be useful in synthesis. Substitution at C-2 and C-6 is tolerated, and the enzyme exhibits only a slight preference for defined stereochemistry at other centers (Table 3.3).

Table 3 Sialic Acids Synthesized by Condensation of Pyruvate with Analogues of ManNAc in the Presence of NeuAc Aldolase

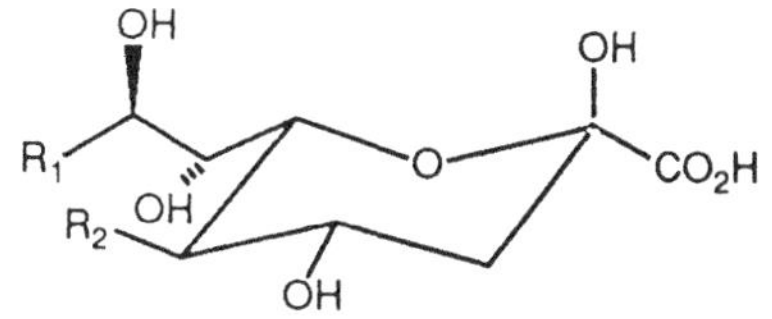

R_1	R_2	Ref.
CH_2OAc	NHAc	97
HCH_2N_3	NHAc	97
CH_2OH	OH	96
CH_2OH	H	93
H	OH	93
CH_2OH	CH_2OH	92
CH_2OH	NHLac	92

In addition to its acceptance of unnatural substrates, several other characteristics make NeuAc aldolase a useful catalyst in synthesis. The cloning of the enzyme has reduced its cost and offers the potential to produce large quantities of proteins with improved stability or with altered stereoselectivity [98]. This approach could be used to extend the chain of a variety of aldoses by two carbon units. Although the optimal pH for activity of NeuAc aldolase is near pH 7.5 at 37°C, the enzyme is active at pH 7–9 [82,83]. The protein is stable in the presence of oxygen and does not require added cofactors [82,83]. One drawback is that an excess of pyruvate (the less expensive reagent) must be used in synthetic reactions to shift the equilibrium toward the formation of product; approximately 7 equivalents of pyruvate is needed to attain 60% conversion of ManNAc to NeuAc at equilibrium. It may be possible to avoid the need for an excess of pyruvate by coupling the synthesis of NeuAc to a more thermodynamically favored process.

An advantage of the enzymatic route compared with the chemical route is that purification of NeuAc or its derivatives may be avoided. For example, an unpurified solution of NeuAc generated enzymatically from an unpurified solution of ManNAc was used in the enzymatic synthesis of cytidine 5'-monophospho-NeuAc (CMP-NeuAc) [89]. The unpurified preparation of ManNAc was derived from base-catalyzed epimerization of the much less expensive starting material, GlcNAc.

TRANSKETOLASE

Transketolase catalyzes the reversible transfer of the hydroxyketo group of a ketose phosphate to an aldose phosphate [99,100]. The cofactor thiamin pyrophosphate (TPP) is associated with the enzyme and activities the ketose (Scheme 7). Most known donor ketoses (xylulose-5-phosphate, sedohep-

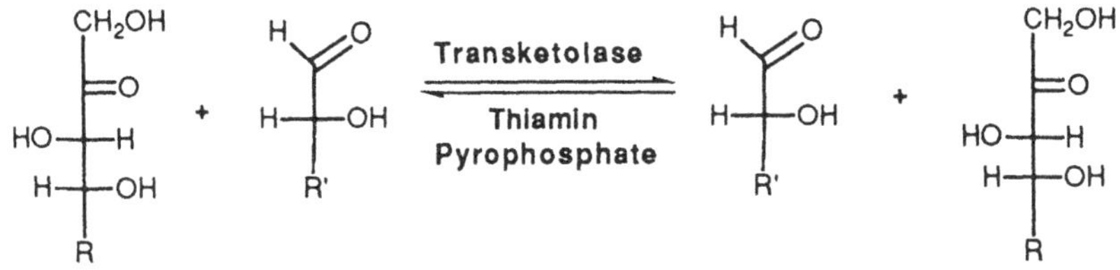

tulose-7-phosphate, fructose-6-phosphate, l-erythrose) have a trans arrangement of hydroxyl groups at C-3 and C-4 in the Fischer projection; hydroxypyruvate is an exception [101]. A range of aldehydes (such as d-glyceraldehyde-3-phosphate, d-ribose-5-phosphate, d-erythrose-4-phosphate, glycolaldehyde) are acceptors [102]. Transketolase has been used in synthesis with its natural substrates and has been used to prepare ^{14}C-labeled intermediates of the pentose pathway [103,104].

Although the substrate specificity of transketolase has not been thoroughly explored, it appears to be a promising catalyst for use in synthesis. The 2-hydroxypyruvate (26) can replace the ketose (25), providing a reactive hydroxyketo group after decarboxylation of hydroxypyruvate. This group is transferred to the acceptor aldehyde in an irreversible reaction (Scheme 8, see page 105) to give (27) [102].

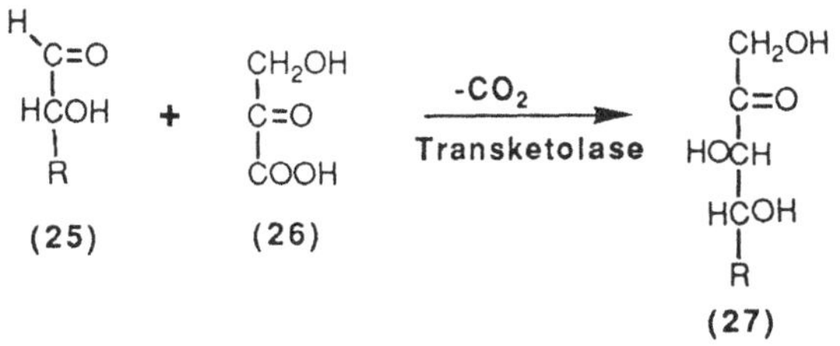

Table 4 Products Synthesized Using Transketolase as a Catalyst and Hydroxy-pyruvate as a Donor

Substrate	Product	Reference
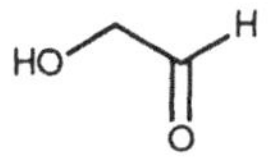	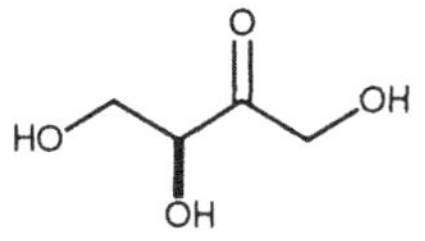	102
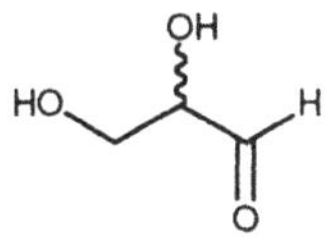	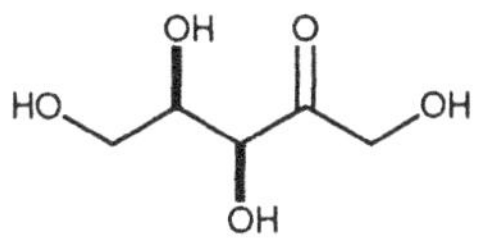	102
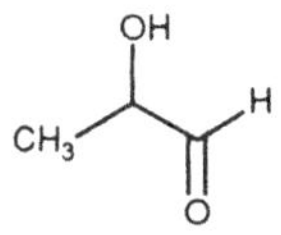	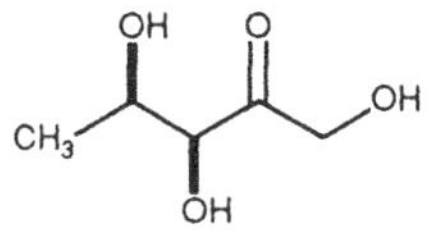	102
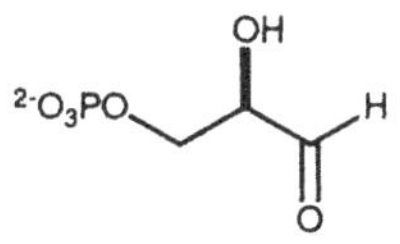	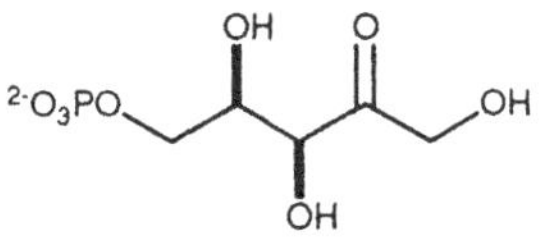	103
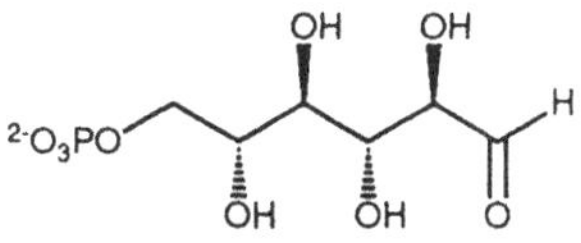	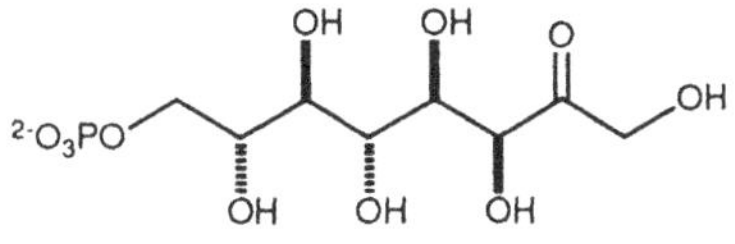	104
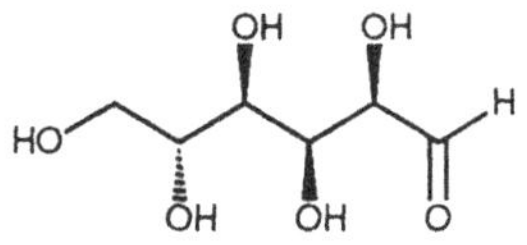	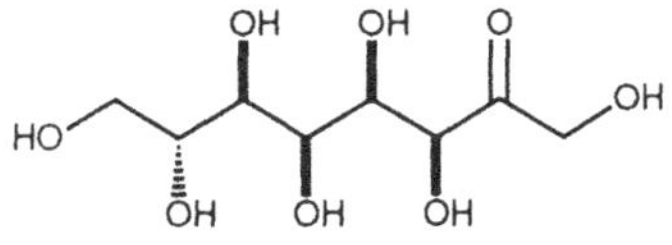	104

This method has allowed the synthesis of a number of monosaccharides on scales of 2–5 mmol with yields from 24 to 85% (Table 4) [102–104]. The enzyme is commercially available and has also been immobilized [102].

KETODEOXYOCTANOATE SYNTHETASE

Ketodeoxyoctanoate (KDO) synthetase catalyzes the reaction of arabinose-5-phosphate (29) (Ara-5-P) and phosphoenolpyruvate (PEP) to form KDO-8-P (30) (Scheme 9; see page 97) [105]. KDO synthetase is not commer-

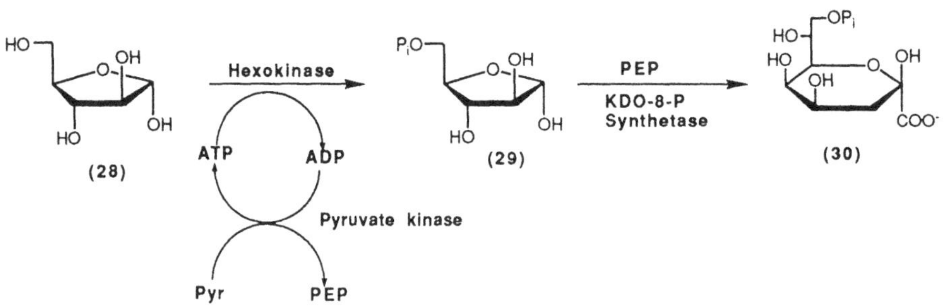

cially available but has been isolated from *E. coli* and used in the synthesis of KDO-8-P (63% from Ara-5-P, 38 mmol [106]. KDO-8-P is a key intermediate in the synthesis of the lipopolysaccharide (LPS) region of Gram-negative bacteria. Inhibitors of LPS biosynthesis are targets for the design of antimicrobial pharmaceuticals [107,108].

The substrate specificity of this enzyme has not been thoroughly examined for synthetic application. In the example given in Scheme 9, the expensive arabinose-5-phosphate was generated from arabinose by hexokinase-catalyzed phosphorylation.

DAHP SYNTHETASE

DHAP synthetase is an enzyme that produces 3-deoxy-d-*arabino*-heptulsonic acid 7-phosphate (32) (DHAP) from d-erythrose-4-phosphate (31). The enzyme was used by Frost [109,110] to synthesize DHAP as an intermediate in the chemical synthesis of its phosphonate analogue 3-deoxy-d-*arabino*-heptulsonic acid 7-phosphate (DAH phosphonate), a potential inhibitor of the shikimate pathway (Scheme 10).

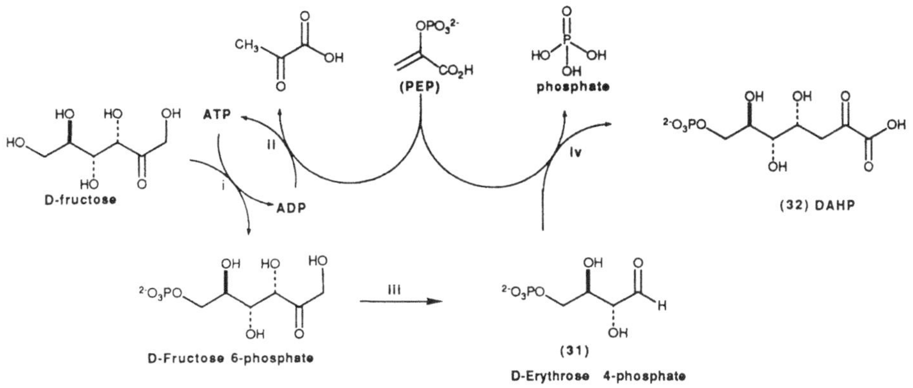

Enzymes: i, hexokinase; ii, pyruvate kinase; iii, transketolase; iv, DAHP synthetase

CONCLUSION AND FUTURE PERSPECTIVES

A large number of other aldolases have been isolated and characterized [33,42,43,111–127]. Limited explorations of substrate specificity have been made in many cases, but these enzymes have not yet been used in synthetic organic chemistry. In terms of potential utility as catalysts, the aldolases of bacterial origin may be of more use than the enzymes from plant or animal sources because the former are more easily cloned and altered by genetic engineering than the later (Table 3.5). The alterations should prove to be useful for controlling the substrate specificity of these enzymes.

Table 5 Summary of Aldolases Available for Use in Asymmetric Organic Synthesis

Nucleophile	Electrophile	Enzyme	Product	Reference
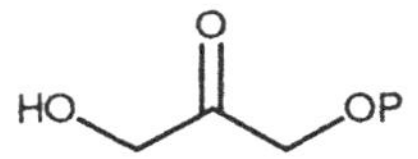	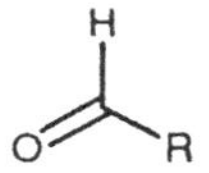	Fructose-1,5-diphosphate aldolase (E.C. 4.1.2.7)	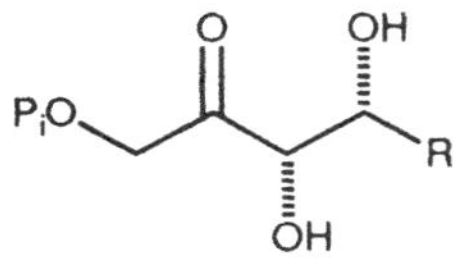	42,43,111
		Fuculose-1-phosphate aldolase (E.C. 4.1.2.17)	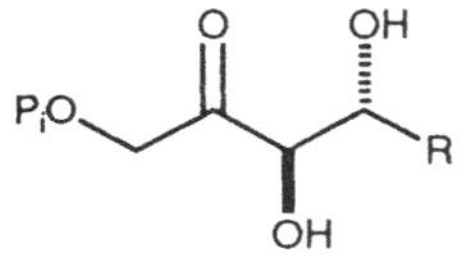	112
		Rhamulose-1-phosphate aldolase (E.C. 4.1.2.19)	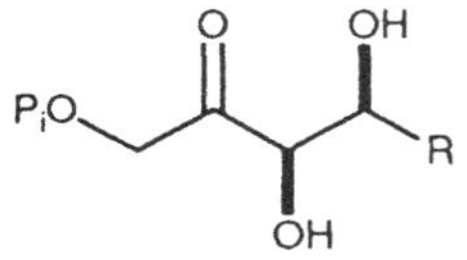	113
		Erythulose-1-phosphate aldolase (4.1.2.2)	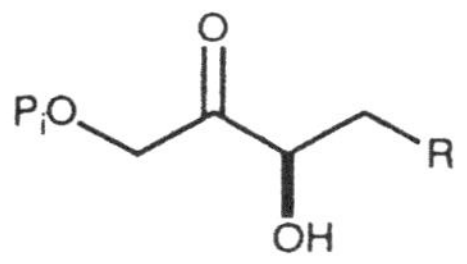	114

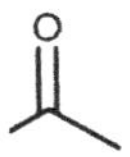

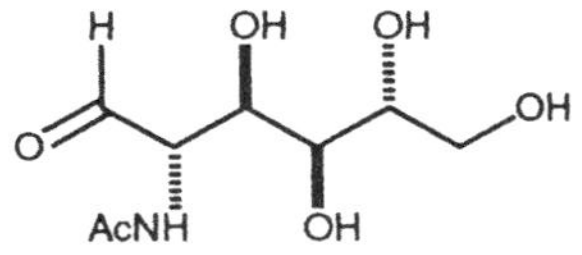

90,115

116

117

117

**Neuraminic acid
Aldolase
(E.C. 4.1.3.3)**

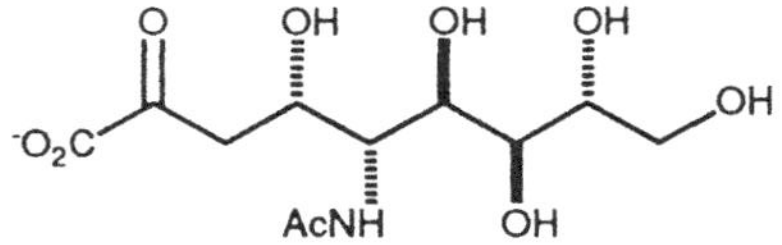

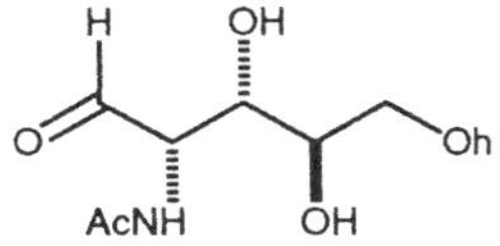

**2-Keto-3-deoxy
octanoate aldolase
(E.C. 4.1.2.23)**

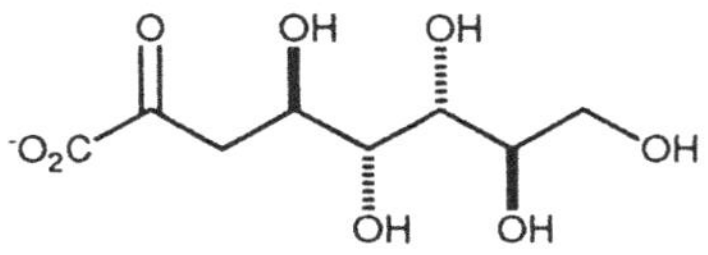

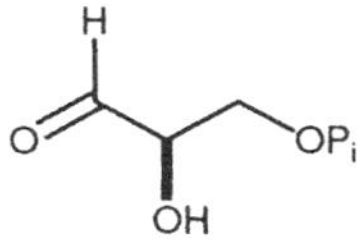

**2-Keto-3-deoxy-6-
phosphogluconate
aldolase
(E.C. 4.1.2.14)**

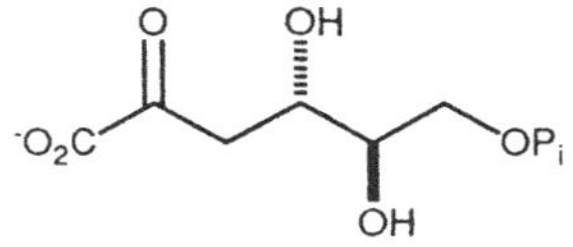

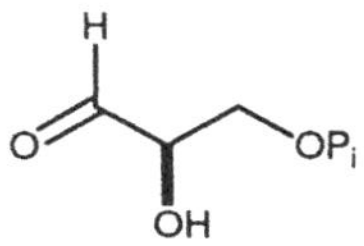

**2-Keto-3-deoxy-6-
phosphogalactonate
aldolase
(E.C. 4.1.2.21)**

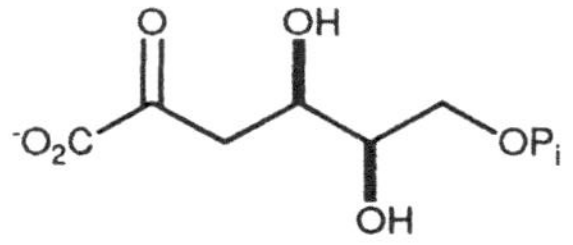

Table 5 Continued

Nucleophile	Electrophile	Enzyme	Product	Reference
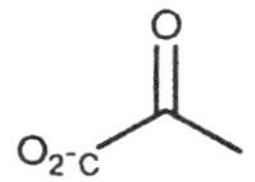 (pyruvate structure, O_2^-C)	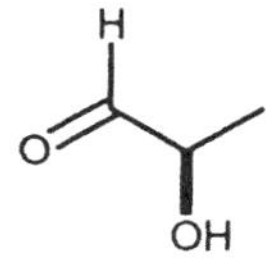 (structure)	2-Keto-3-dexoy-L-arabinoate aldolase (E.C. 4.1.2.18)	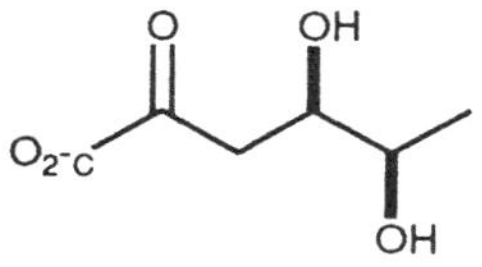 (structure)	117
	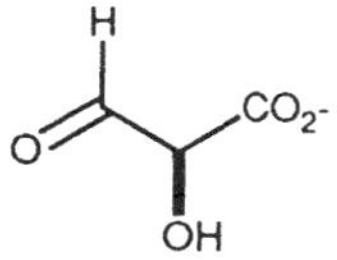 (CO_2^- structure)	2-Keto-3-deoxy-D-glucoarate aldolase (E.C. 4.1.2.20)	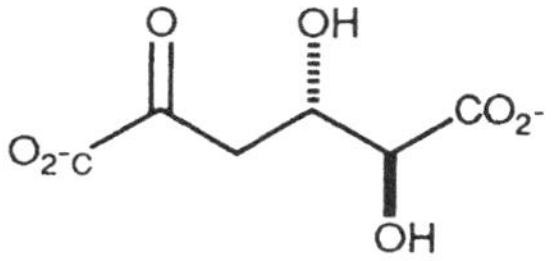 (structure, CO_2^-)	117
	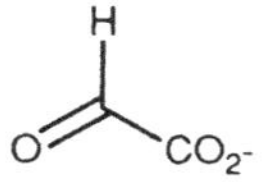 (CO_2^- structure)	2-Keto-4-hydroxy-glutarate aldolase (E.C. 4.1.2.31)	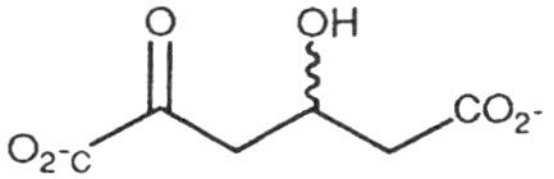 (structure, CO_2^-)	117
	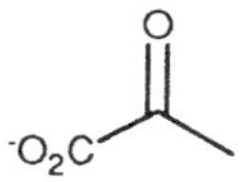 (acetone, ^-O_2C)	4-Methyl-4-hydroxy-2-ketoglutarate aldolase	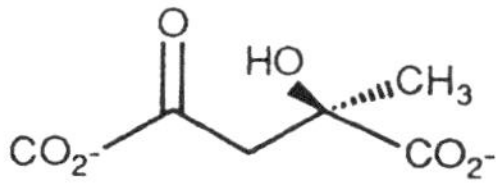 (structure, CH_3, CO_2^-, CO_2^-)	117

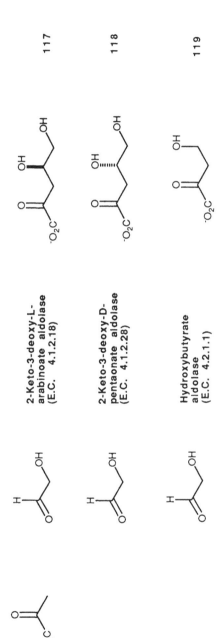

117

118

119

2-Keto-3-deoxy-L-
arabinoate aldolase
(E.C. 4.1.2.18)

2-Keto-3-deoxy-D-
pentaonate aldolase
(E.C. 4.1.2.28)

Hydroxybutyrate
aldolase
(E.C. 4.2.1.1)

Table 5   Continued

Nucleophile	Electrophile	Enzyme	Product	Reference
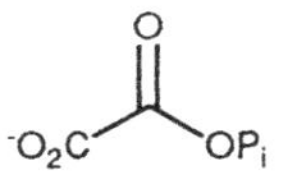	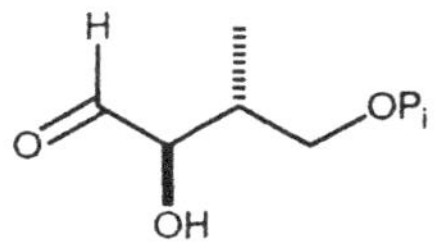	Phosphono-2-keto-3-deoxyheptanoate aldolase (E.C. 4.2.1.15)	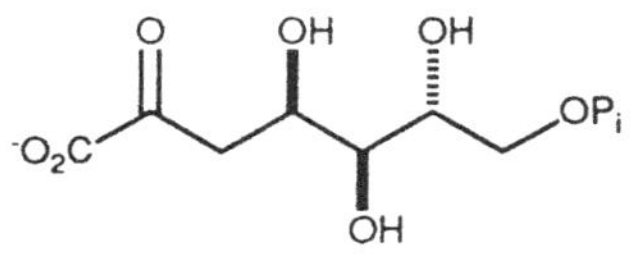	120
	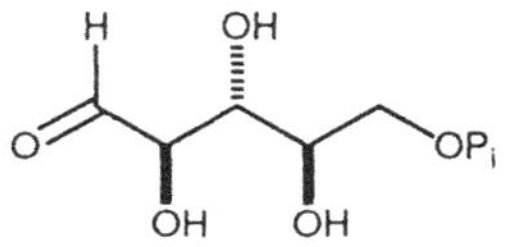	Phosphono-2-keto-3-deoxyoctanoate (E.C. 4.2.1.16)	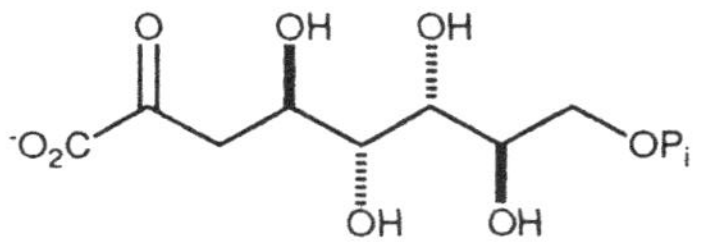	121

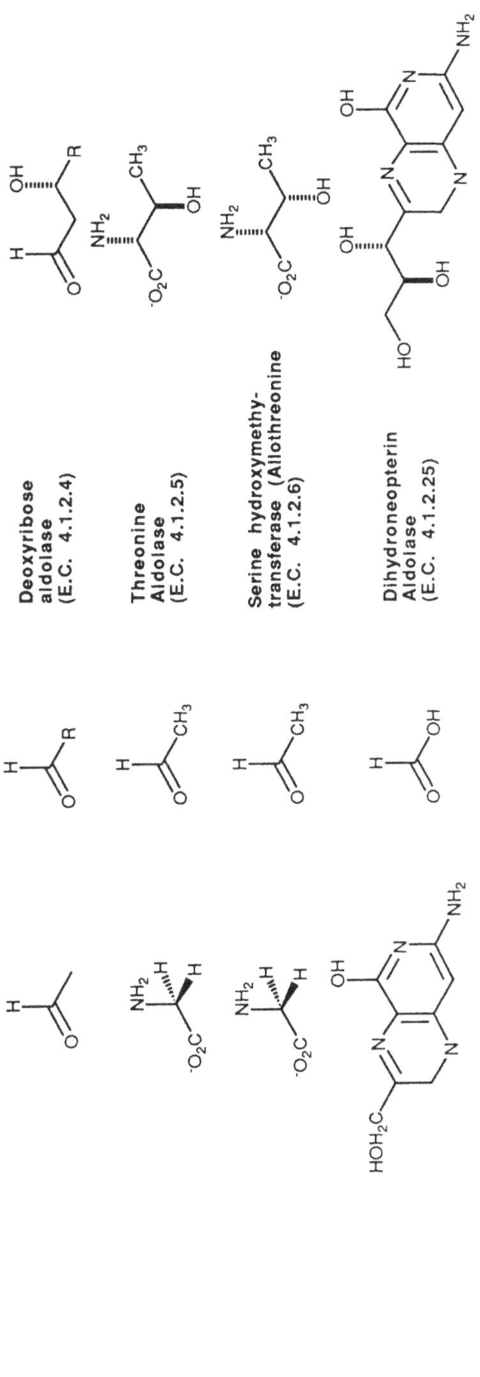

122

123

124

125

Deoxyribose
aldolase
(E.C. 4.1.2.4)

Threonine
Aldolase
(E.C. 4.1.2.5)

Serine hydroxymethy-
transferase (Allothreonine
(E.C. 4.1.2.6)

Dihydroneopterin
Aldolase
(E.C. 4.1.2.25)

REFERENCES

1. Bartmann, W., and Sharpless, K. D. (eds.). *Stereochemistry of Organic and Bioorganic Transformations*, VCH, New York, 1987.
2. Braun, M. *Agnew. Chem., Int. Ed. Engl., 26*, 24 (1987).
3. Reetz, M. S. *Agnew. Chem., Int. Ed. Engl. 23*, 556 (1984).
4. Morrison, J. D., (ed.). *Asymmetric Synthesis,* Academic Press, New York, Vol. 2, 1983; Vol. 35 (1984).
5. Eliel, L., and Otsuka, S. (eds.), *Asymmetric Reactions and Processes in Chemistry*, ACS Symposium Series 185, American Chemical Society, Washington, DC, 1982.
6. Trost, B. M., and Hutchinson, C. R. (eds.), *Organic Synthesis Today and Tomorrow*, Pergamon Press, New York, 1981.
7. Evans, D. A., Nelson, J. V. and Taber T. R. *Top Stereochem. 13*, 1 (1982).
8. Masamune, S., Choy, W., Petersen, J. S., and Sita, J. R. *Agnew. Chem., Int. Ed. Engl., 24*, 1 (1985).
9. Heathcock, C. H. In *Comprehensive Carbanion Chemistry*, Vol. 2 (E. Buncel and T. Durst, eds.), Elsevier, Amsterdam, 1984, p. 177.
10. Meyers, A. I. *Aldrichchim. Acta, 18*, 59 (1985).
11. Oppolzer, W. *Agnew. Chem., Int. Ed. Engl. 23*, 876 (1984).
12. Helmchen, G., and Wierzchowski, R. *Agnew. Chem., Int. Ed. Engl. 23*, 60 (1984).
13. Seebach, D, and Prelog, V. *Agnew. Chem., Ed. Engl. 21*, 654 (1982).
14. Enders, D., Fey, P., and Kipphardt, H. *Org. Prep. Proced. Int. 17*, 1 (1985).
15. Mukaiyama, T. *Org. React. 28*, 203 (1982).
16. Danishefsky has reported a catalytic asymmetric hetero-Diels–Alder reaction using Eu(hfc)₃ that gives aldo-type products, albeit in only moderate enantiofacial selectivity (58% enantiomeric excess): M. Bednarski, C. Maring, and S. Danishefsy, *Tetrahedron Lett. 33*, 3451 (1983). Chiral amine complexes have been used to give enantiomerically enriched products: A. Ando and T. Shioiri, *J. Chem. Soc., Chem. Commun.* 1620 (1987).
17. Nakagawa, M., Nakao, H., and Watanabe, K. -i, *Chem. Lett.* 391 (1985).
18. Mikami, K., Terada, M., and Nakai, T. Paper 3Y29, presented at the 52nd Annual Meeting of the Chemical Society of Japan, Kyoto, April 1–4 1986.
19. Ito, Y., Sawamura, M., and Hayashi, T. *J. Am. Chem. Soc. 108*, 6405 (1986).
20. Whitesides, G. M., and Wong, C. H. *Agnew. Chem., Int. Ed. Engl. 24*, 617 (1985).
21. Jones, J. B. *Tetrahedron, 42*, 3351 (1986).
22. Akiyama, A., Bednarski, M. D., Kim, M. -J., Simon, E. S., Waldmann, H., and Whitesides, G. M., *Chem. Br. 23*, 645 (1987).
23. Porter, R., and Clark, S. (eds.), *Enzymes in Organic Synthesis: Ciba Foundation Symposium III*, Pitman, London, 1985.
24. Whitesides, G. M., and Wong, C. -H. *Aldrichchim. Acta, 16*, 27 (1983).
25. Jones, J. B. In *Asymmetric Synthesis*, Vol. 5 (J. D. Morrison, ed.), Academic Press, New York, 1985, p. 309.

26. Suckling, C. J., and Wood, H. C. S. *Chem. Br. 15*, 243 (1979).
27. Wong, C. -H. In *Enzymes as Catalysts in Organic Synthesis* (M. P. Schneider, ed.), Reidel, Dordrecht, 1986, p. 199.
28. Borgstrom, B., and Brockman, H. L. (eds.), *Lipases*, Elsevier, Amsterdam, 1984.
29. Jones, J. B., and Beck, J. F. In *Techniques in Chemistry* (J. B. Jones, C. S. Sih, and D. Perlman, eds), Wiley-Interscience, New York, 1976, p. 107.
30. Kim, M. -J., and Whitesides, G. M. *J. Am. Chem. Soc. 110*, 2959 (1988).
31. Simon, E. S., Plante, R., and Whitesides, G. M., *Appl. Biochem. Biotechnol. 22*, 169 (1989).
32. Chenault, H. K., Dahmer, J., and Whitesides, G. M. *J. Am. Chem. Soc. 111*, 6354 (1989).
33. Bednarski, M. D., Simon, E. S., Bischofberger, N., Fessner, W. -D., Kim, M. -J., Lees, W., Saito, T., Waldmann, H., and Whitesides, G. M. *J. Am. Chem. Soc., 111*, 628 (1989).
34. Durrwatcher, J. R., and Wong, C. -H. *J. Org. Chem. 53*, 4175 (1988).
35. Mosbach, K. (ed.), *Methods in Enzymology*, Vol. 44 Academic Press, New York, 1976.
36. Moshbach, K. (ed.), *Methods in Enzymology*, Vol. 136 Academic Press, New York, 1987.
37. Chibata, I. *Immobilized Enzymes—Research and Development*, Halsted Press, New York, 1978.
38. Lee, E. K. In *Encyclopedia in Physical Science and Technology*, Vol. 8, Academic Press, New York, 1987, p. 20.
39. Bednarski, M. D., Chenault, K. H., Simon, E. S., and Whitesides, G. M. *J. Am. Chem. Soc. 109*, 1283 (1987).
40. Chenault, K. H., and Whitesides, G. M. *Appl. Biochem. Biotechnol. 14*, 147 (1987).
41. Chenault, K. H., Simon, E. S., and Whitesides, G. M., *Biotechnol. Genet. Eng. Rev.* (Russell, G. E., ed.), Intercept: Wimborne, Dorset, 1988, Vol. 6, Chapter 6.
42. Serianni, A. S., Cadman, E., Pierce, J., Hayes, M. L., and Barker, R. *Methods Enzymol. 89*, 83 (1982).
43. Horecker, B. L., Tsolas, O., and C. Y. Lai, In *The Enzymes*, Vol. 7, 3rd ed. (P. D. Boyer, ed.), Academic Press, New York, 1972, p. 213.
44. Morse, D. F., and Horecker, B. L. In *Advances in Enzymology*, Vol. 31, (F. F. Nord, ed.), Wiley-Interscience, New York, 1968, p. 125.
45. Willnow, P. In *Methods of Enzymatic Analysis*, 3rd ed. (H. U. Bergmeyer, J. Bergmeyer, and M. Grassl, eds), Verlag Chemie, Weinheim, 1984: Vol. II, p. 146, Vol. IV, p. 346.
46. Barman, T. E. (ed.), *The Enzyme Handbook*, Vol. II, Springer-Verlag, New York, 1969, P. 736.
47. Bergmeyer, H. U., Bergmeyer, J. and Grassl, M. (eds.). *Methods of Enzymatic Analysis*, Vol. IV, Verlag Chemie, Weinheim, 1984, p. 421.
48. Racker, E. In *The Enzymes*, Vol V, 3rd ed (P. D. Boyer, ed.), Academic Press, New York, 1961, p. 397.

49. Durrwachter, J. R., Grueckhammer, D. G., Nazaki, K., Sweers, H. M., and Worg, C. -H. *J. Am. Chem. Soc. 108*, 7812 (1986).
50. Wong, C. -H., and Whitesides, Gi. M. *J. Org. Chem. 48*, 3199 (1983).
51. Wong, C. -H., Mazenod, F. P., and Whitesides, G. M. *J. Org. Chem. 48*, 3493 (1983).
52. Bednarski, M. D., Waldmann, H. J., and Whitesides, G. M. *Tetrahedron Lett.*, 5807 (1986).
53. Jones, J. K. N., and Sephton, H. H. *Can J. Chem.*, 753 (1960).
54. Hough, L., and Jones, J.l K. N. *J. Chem. Soc.* 4047 (1952).
55. Jones, J. K. N., and Matheson, N. K. *Can. J. Chem. 37*, 1754 (1959).
56. Hough, L., and Jones, J. K. N. *J. Chem. Soc.* 342 (1953).
57. Haustveit, G. *Carbohydr. Res. 47*, 164 (1976).
58. Franke, F. P., Kapuscinski, M., MacLeod, J. K., and Williams, J. F. *Carbohydr. Res. 125*, 177 (1984).
59. Mehler, A. H., and Cusic, M. E., Jr. *Science, 155*, 1101 (1967).
60. Gorin, P. A. J., and Jones, J. K. N. *J. Chem. Soc.* 1537 (1953).
61. Gorin, P. A. J., Hough, L., and Jones, J. K. N. *J. Chem. Soc.* 3843 (1955).
62. Jones, J. K. N., and Kelly, R. B. *Can. J. Chem. 34*, 95 (1956).
63. Kapuscinski, M., Franke, F. P., Flanigan, I. Macleod, J. K., and Williams, J. F. *Carbohydr. Res. 140*, 65 (1985).
64. Bally, C., and Leuthardt, F. *Helv. Chim. Acta, 53* (1970).
65. Horecker, B. L., and Smyrniotis, P. Z. *J. Am. Chem. Soc. 74*, 2133 (1952).
66. Effenberger, F., and Straub, A. *Tetrahedron Lett. 28*, 1641 (1987).
67. Turner, N. J., and Whitesides, G. M. *J. Am. Chem. 111*, 624 (1989).
68. Huang, P. C., and Miller, O. N. *J. Biol Chem. 230*, 805 (1958).
69. Hough, L., and Jones, J. K. N., *J. Chem. Soc.* 4052 (1952).
70. Byrne, W. L., and Lardy, H. A. *Biochem. Biophys. Acta, 14*, 495 (1954).
71. Smyrnigotis, P. Z., and Horecker, B. L. *J. Biol. Chem. 187*, 325 (1950).
72. Lehninger, A. L., and Sice, J. *J. Am. Chem. Soc. 77*, 5343 (1955).
73. Ballou, C. E., Fischer, H. O. L., and MacDonald, D. L. *J. Am. Chem. Soc., 77*, 5967 (1955).
74. Charalampous, F. C. *J. Biol. Chem. 211*, 249 (1954).
75. Gorin, P. A. J., Hough, L., and Jones, J. K. N. *J. Chem. Soc.* 2140 (1953).
76. Crans, D. C., and Whitesides, G. M. *J. Am. Chem. Soc. 107*, 7019 (1985).
77. Goldberg, M., Fessenden, J. M., and Racker, E. In *Met in Enzymology*, Vol. 9 (W. A. Wood, ed), Academic Press, New York, (1966), p. 515.
78. Ziegler, T., Straub, A., and Effenberger, F. *Agnew. Chem., Int. Ed. Engl. 27*, 716 (1988).
79. Pederson, R. L. Kim, M. -J., and Wong, C. -H. *Tetrahedron Lett. 29*, 4645 (1988).
80. Brunetti, P., Jourdian, G. W., and Roseman, S. *J. Biol. Chem. 237*, 2447 (1962).
81. DeVries, G. H., and Brinkley, S. B. *Archiv. Biochem. Biophys. 151*, 234 (1972).
82. Uchida, Y., Tsudaka Y., and Sugimori, T. *J. Biol. Chem. 96*, 507 (1984).
83. Uchida, Y. Tsudaka, Y., and Sugimori, T. *Agric. Biol. Chem. 96*, 181 (1985).
84. Schauer, R. *Trends Biochem Sci. 337*, 557 (1985).

85. See Introduction, Ref. 8, Chapter 8.
86. Corfield, A. P., and Schauer, R. In *Sialic Acids* (R. Schauer, ed.), Springer-Verlag, New York, 1982, pp. 195 ff.
87. McGuire, E. J. In *Biological Roles of Sialic Acid* (A. Rosenberg and C. -L. Schengrund, eds.), Plenum Press, New York, 1976, Chapter 4.
88. Gottschalk, A. *The Chemistryf and Biology of Sialic Acids and Related Substances*, Cambridge University Press, Cambridge, 1960.
89. Simon, E. S., Bednarski, M. D., and Whitesides, G. M. *J. Am. Chem. Soc. 110,* 7159 (1988).
90. Auge, C., and Gautheron, C. *Tetrahedron Lett. 25,* 4663 (1984).
91. Kim, M. -J., Hennen, W. J., Sweers, H. M., and Wong, C. -H. *J. Am. Chem. Soc. 110,* 1260 (1988).
92. Suttajit, M., Urban, C., and McLean, R. L. *J. Biol. Chem. 246,* 810 (1971).
93. Schauer, R., Wember, M., Wirth-Peitz, F., and Do Araral, C. F. *Hoppe-Seylers Z. Phsiol. Chem. 352,* 1073 (1971).
94. Faillard, V. H., Do Amaral, C. F., and Blohn, M. *Hoppe-Seylers Z. Physiol. Chem. 350,* 798 (1969).
95a. Augé, C., and Gautheron, C. *J. Chem. Soc., Chem. Comm.* 859 (1987).
95b. Augé, C., *New J. Chem. 12,* 733 (1988).
96a. Augé, C., et al., *Tetrahedron Lett. 26,* 2439 (1985).
96b. Auge, C., et al., *Tetrahedron Lett. 30,* 2217 (1989).
97. Brossmer, R. Rose, U., Kasper, D., Smith, T. L., Grasmuk, H., and Unger, F. M. *Biochem. Biophys. Res. Commun., 96,* 1282 (1980).
98. Ohta, Y., Shimosaka, M., Murata, K., Tsudaka, Y., and Kimura, A. *Appl. Microbiol. Biotechnol. 98,* 1216 (1986).
99. Racker, E. In *The Enzymes* (P. D. Boyer, H. Lardy, and K. Myrbach, eds.), Academic Press, New York, 1961, p. 397.
100. Kochetov, G. A. *Methods Enzymol. 90,* 209 (1982).
101. Kochetov, G. A., and Solov'eva, O. N. *Biokhimiya (Moscow) 42,* 1872 (1977). *Chem. Abstr. 87,* 196453 p.
102. Bolte, J., Demuynck, C., and Samaki, H. *Tetrahedron Lett. 28,* 5525 (1987).
103. Arora, K. K., MacLeod, J. K., and Williams, J. F. *J. Labelled Compd. Radiopharm., 24,* 205 (1987).
104. Mocali, A., Aldinucci, D., and Paoletti, F. *Carbohydr. Res. 143,* 288 (1985).
105. Ray, P. H. *J. Bacteriol. 141,* 635 (1980).
106. Bednarski, M. D., Crans, D. C., DiCosmio, R., Simon, E. S., Stein, P. D., Schnieder, M., and Whitesides, G. M. *Tetrahedron Lett. 29,* 427 (1988).
107. Hammond, S. M., Claesson, A. Jansson, A. M., Larsson, L. - G., Pring, B. G., Town, C. M., and Ekström, B. *Nature, 327,* 730 (1987); Luthman, K., Orve, M., Waglund, T., and Claesson, A. *J Org Chem. 52,* 3777 (1987).
108. Danishefsky, and DiNinno, M. P. *Agnew. Chem. Int. Ed. Engl., 26,* 15 (1987).
109. Frost, J. W., and Knowles, J. R. *Biochemistry, 23,* 4465 (1984).
110. Reiner, L. M., Donley, D. L. Pompliano, D. L. and Frost, J. W. *J. Am. Chem. Soc., 108,* 8010 (1986).

111. Rose, I. A. *J. Am. Chem. Soc. 80*, 5835 (1958).
112. Ghalambar, M. A., and Heath E. C. *J. Biol. Chem. 237*, 2427 (1962).
113. Chin, T. H., and Geingold, D. S. *Biochemistry, 8,* 98 (1969).
114. Charllampous, G. C. *Methods Enzymol., 5*, 283 (1962).
115. Comb, D. G., and Roseman, S. *J. Biol. Chem. 235*, 2595 (1960).
116. Ghalambar, M. A., and Heath, E. C. *J. Biol. Chem. 241*, 3222 (1966).
117. Wood, W. A. *In the Enzymes*, Vol. III (P. D. Boyer, Ed), Academic Press, New York, 1970, p. 281.
118. Dahma, A. S. *Biochem. Biophys. Res. Commun. 60*, 1433 (1974).
119. Hifgt, H., and Mahler, H. R. *J. Biol. Chem. 198,* 901 (1952).
120. Weissbach, A., and Hurwitz, J. *J. Biol. Chem. 234*, 750 (1959).
121. Ray, P. H. *J. Bacteriol., 141*, 635 (1980).
122. Hoffee, P., Rosen, O. M., and Horecker, B. L. *Methods Enzymol. 9,* 545 (1966).
123. Malkin, L. I., and Greenber, D. M. *Biochim. Biophys. Acta, 85*, 117 (1964).
124. Greenber, D. M. *Methods Enzymol. 5*, 931 (1962).
125. Mathis, J. B., and Brown, G. M. *J. Biol. Chem. 245*, 3015 (1970).
126. Heath, E. C., Hurwitz, J., Horecker, B. L., and Ginsberg, S. *J. Biol. Chem. 231*, 1009 (1958).
127. Racker, E. *Methods Enzymol. 5*, 276 (1962).

4

Enzymatic Synthesis of Complex Carbohydrates and Their Glycosides

Kurt G. I. Nilsson
Chemical Center, University of Lund, Lund, Sweden

INTRODUCTION

The importance of the complex carbohydrate structures in a variety of biological processes is now well recognized [1-6]. It has long been appreciated that there is extensive glycosylation of lipids and proteins at the extracellular surface. However, detailed knowledge of the structure and function of glycolipids and glycoprotein carbohydrates has largely developed over the past two decades. More recently, it has become clear that several proteins within the cell are also conjugated to carbohydrates [7].

Beyond their roles as blood group determinants, carbohydrate structures have been found to be important for the secretion, activity, immunogenicity, and circulation half-life of glycoproteins (including recombinant proteins) [8,9], in cell–cell interactions (e.g., fertilization, embryogenesis, organ development, cell migration), and as cell surface receptors for toxins and pathogens [1-6]. Several cancer-associated carbohydrate structures are known, and the medical importance of monoclonal antibodies to these antigens in diagnosis and therapy is being evaluated [2,5,10-12].

The availability of the relatively short glycoconjugate oligosaccharides (often < 20 sugar residues) and of shorter biologically active fragments, as well as their analogues, in amounts sufficient for application and fundamental

117

study, is important. The structures available today are mainly produced by classical chemical synthesis or by isolation from biological materials. Classical chemical methods are well developed but require multistep procedures for selective synthesis [13,14]. Preparative isolation of the structures of interest from the complex mixtures of carbohydrates present in low concentrations in biological material is cumbersome.

Thus, there is a need for simple methods suitable for production of glycoconjugate carbohydrates. Enzymes are attractive in this field because they catalyze stereo- and regioselective synthesis of complex structures with a minimum of reaction steps, under mild conditions, and in aqueous solutions. Enzymatic methods are the most plausible for modification of carbohydrates on proteins and cell surfaces. This chapter evaluates the potential of enzymatic in vitro synthesis of mammalian glycoconjugate carbohydrate structures. The glycoconjugate field is briefly discussed. Proteoglycan carbohydrates (heparin, chondroitin sulfate, etc.) are not included.

Glycoconjugate Carbohydrates: Structure, Function, and Applications

Structure

The oligosaccharide structures of glycoconjugates seldom contain more than 20 residues and are built from relatively few glycosidically linked monosaccharides (L-fucose, D-galactose, D-mannose, N-acetyl-D-glucosamine, N-acetyl-D-galactosamine, and N-acetylneuraminic (sialic) acid, glucose, and xylose) [1,2]. Still, because of the many possible linkage sites and alternative anomeric configuration (α- or β-glycosidic linkages), a huge number of structures can be generated (cf. Fig. 4.1). Thus, four monosaccharides can join

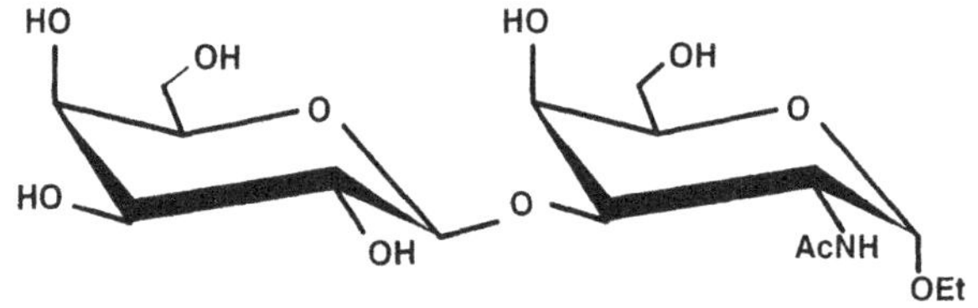

Galβ 1-3GalNAcα-OEt

Figure 4.1 Ethyl 2-acetamido-2-deoxy-3-O-β-D-galactopyranosyl-α-D-galactopyranoside; the α-ethyl glycoside of the T-antigen [5].

to more than 35,000 distinct tetrasaccharides. Further diversification may occur by covalent attachment of acetyl, phosphate, or sulfate groups to the sugars.

In oligopeptides or oligonucleotides, only the sequence and number of monomeric units determine the number of different structures; carbohydrates, in comparison have a much higher potential for encoding biological information, and it has been proposed that "the specificity of many natural polymers is written in terms of sugar residues and not of amino acids or nucleotides" [4]. Moreover, the oligosaccharide chains have been shown to adopt solution conformations in which well-defined regions with hydrophobic and hydrophilic groups are exposed [15-17]. Such regions have been proposed to serve as specific binding sites for proteins.

Structural analysis employing enzymes, permethylation analysis, and NMR spectroscopy has revealed a limited number of carbohydrate core structures for glycolipids and for the N- and O-linked glycoproteins. The glycolipid cores may be repeated or extended, and finally ended with blood group determinants and other terminal glycosylation sequences (Table 4.1)[18]. The three main N-linked (Asn-linked) carbohydrate types all contain the common Man_3 $GlcNAc_2$ core, and the five O-linked core structures share the GalNAc-OThr/ Ser structure (Table 4.2). Diversity of N-glycoprotein carbohydrate structures is achieved by the extension of core structures (e.g., with $Gal\beta1$-$4GlcNAc\beta1$-3), "branching" (i.e., the formation of bi-, tri-, and higher antennary structures), and by different terminal sequences (e.g., blood group structures or sialylated structures).

The glycosylation of lipids and proteins is species- and tissue-specific, and is thought to be largely determined by the differential expression of the various glycosyltransferase genes [1,3]. The importance of species-specific N-glycosylation of recombinant proteins has been recognized [19-22]. Yeast cells produce glycoproteins, which may exert bioactivity, but the presence of terminal mannose sequences leads to their rapid clearance from the circulation by specific mannose receptors on macrophages and liver hepatocytes [8]. Asialo- (Gal-terminated) and asialoagalacto (GlcNAc-terminated) glycoproteins are also rapidly cleared from the circulation by specific receptors [8].

Mammalian cell lines have been used to prolong the circulation half-life and decrease the immunogenicity of recombinant proteins. However, with these cell lines the terminal sequences (e.g., $Neu5Ac\alpha2$-3 and $Neu5Ac\alpha2$-6 linkages), the ratios of branched structures, and the amount of $Gal\beta1$-4GlcNAc repeats may be changed compared with the "natural glycoform" [19-22]. The use of glycosidase and glycosyltransferase for modification of recombinant glycoproteins has been reported [23].

Table 4.1 Structures of Some Glycosphingolipids

Glycolipid	Name (abbreviations)
Ganglio series	
Neu5Acα2–3Galβ1–3GalNAcβ1–4Galβ1–4Glcβ1–1Cer	GM$_{1b}$
Neu5Acα2–8Neu5Acα2–3Galβ1–4Glcβ1–1Cer	GD$_2$
$\qquad\qquad\qquad\quad$ 4	
$\qquad\qquad\qquad\quad$ |	
$\qquad\qquad$ GalNAcβ1–4Galβ1	
Globo series	
GalNAcβ1–3Galα1–4Galβ1–4Glcβ1–1Cer	Globo
Lacto series	
Fucα1–2Galβ1–4GlcNAcβ1–3Galβ1–4Glcβ1–1Cer	Lactofucopentaosyl [V] Cer
Muco series	
GalNAcα1–3Galβ1–3Galβ1–4Glcβ1–1Cer	
Gala series	
Galα1–4Galβ1–1Cer	Galabiose

Source: Ref. 18.

Table 4.2 Examples of N-Linked and O-Linked Carbohydrate Sequences of Glycoproteins [1,3,49]

N-Linked Sequences

Oligomannosidic type

```
Manα1-2Manα1-2Manα1
                    |
                    6
                  Manα1
                  3   |
                  |   6
    Manα1-2Manα1      Manβ1-4GlcNAcβ1-4GlcNAcβAsn
                  3
                  |
           Manα1-2Manα1
```

Complex (N-acetyllactosamine) type

```
Neu5Acα2-6Galβ1-4GlcNAcβ1-2Manα1
                                |
                                6
                                  Manβ1-4GlcNAcβ1-4GlcNAcβAsn
                                3
                                |
Neu5Acα2-6Galβ1-4GlcNAcβ1-2Manα1
```

Hybrid type

```
     Neu5Acα2-6Galβ1-4GlcNAcβ1-2Manα1
                                    |
                                    6
Neu5Acα2-6Galβ1-4GlcNAcβ1-2Manα1     Manβ1-4GlcNAβ1-4GlcNAcβAsn
                              |      3
                              4      |
                            Manα1
                            2
                            |
              Neu5Acα2-3Galβ1-4GlcNAcβ1
```

O-Linked Sequences

```
              Neu5Acα2
                  |
                  6
Neu5Acα2-3Galβ1-3GalNAcαThr/Ser
GalNAcα1-3(Galβ1-4GlcNAcβ1-3)₂Galᵦ1
                              |
                              3
                        GalNAcαThr/Ser
                        6
                        |
        Fucα1-2Galβ1-4GlcNAcβ1
```

Function

The biological activity of glycoproteins may be modulated by glycosylation [3]. Examples are the plasma glycoproteins antithrombin III [24], tissue plasminogen activator (tPA) [3], and fibrinogen [25]. The nonglycosylated form of the cell-surface-bound receptor for transferrin has been reported to show a much lower affinity for transferrin than the glycosylated form [26]. Both "up" and "down" regulation by glycosylation of the activity of other glycoproteins have been observed [3].

The importance of cell-bound carbohydrates for cell–cell interactions are well documented, and specific carbohydrate structures that inhibit sperm–egg interaction [27] and embryogenesis [28] have been identified. Several examples of carbohydrate-mediated (or modulated) cell–cell interactions in developing and in fully differentiated cells are known, such as the homing of lymphocytes from the circulation into the lymphoid organs [4]. The homing of lymphocytes is mediated by a cell surface receptor that has been cloned and found to contain a carbohydrate-binding (lectin) domain [29]. Several other intracellular and extracellular lectins have been identified [3,30].

Table 4.3 Cancer–Associated Carbohydrate Antigens [5,10]

Structure	Name	Symptom
Galα1–4Galβ1–4Glc	Gb$_3$	Burkitt's lymphoma
Galβ1–4(Fucα1–3)GlcNAc	Le^x	Colorectal cancer
[Galβ1–4(Fucα1–3)GlcNAcβ1–3]$_3$Gal	poly-Le^x	Human adeno-carcinoma
Galβ1–3GalNAcα1–R	T antigen	Several cancers
GalNAcα1–3GalNAcβ1–3Gal	Forssman antigen	Tumors from Forssman-negative patients
GalNAcα1–3Galβ–GlcβNAc	Blood group A–like antigen	Blood group B and O individuals
Neu5Acα2–8Neu5Acα2–3Galβ1–4Glc	GD$_3$	Human melanoma
Neu5Acα2–3Galβ1–3(Fucα1–4)–GlcNAc	Sia-Le^a	Colon and pancreas cancer

Specific functions for membrane glycoplipids that have been proposed include modulation of cell proliferation through growth factor receptors e.g., GM3 on epidermal growth factor (EGF) receptor tyrosine kinase [31].

Besides the inherited disorders of glycosylation, several diseases have been shown to be accompanied by alterations in glycosylation, including alcoholism (increased levels of asialotransferrin) [32], rheumatoid arthritis (changed glycosylation of IgG leading to aggregation of IgG) [3], and various cancers [5,10].

A number of cancer-associated carbohydrate structures have been identified, and a few are listed in Table 4.3. Some of the structures are related to blood group antigens and are shared by both cell-surface-bound glycolipids and glycoproteins, as well as secreted mucins (O–linked glycoproteins). Several cancer-associated carbohydrate antigens have been found to be developmentally regulated (oncodevelopmental antigens) [5,33]. The metastatic potential of cancer cells has been shown to be correlated with glycosylation changes (e.g., extensive sialylation and GlcNAcβ1-6Man branching) [34].

The importance of cell surface carbohydrates as receptors for toxins, blood group and tumor associated antibodies, and attachment proteins of bacteria, mycoplasma, and viruses is well-documented [5,6,10,11,35]. Representative examples are shown in Tables 4.3, 4.4, and 4.5.

Applications

As shown in Tables 4.3–4.5, short fragments of the glycoconjugate glycans are often sufficient for biological activity. Such structures and their analogues or glycosides have been used in a number of applications [36–46]:

> Affinity chromatography (lectins, antibodies, glycosyltransferases)
> Preparation of neoglycoproteins (e.g., for immunization)
> Incorporation into liposomes
> Targeting of drugs
> Characterization, localization, and assay of antibodies, glycosidases, glycosyltransferases, lectins
> Inhibition or change of glycosylation, via inhibition of glycosyltransferases or glycosidases
> Diagnosis and therapy of diseases (e.g., cancer and infections)

Some *E. coli* strains common in urinary tract infections contain fimbriae with attachment proteins ("adhesins") specific for Galα1-4Gal (blood group P-structure). Recently, a sensitive and rapid assay for the specific identification of P-fimbriated *E. coli* based on the Galα1-4Gal receptor structure was described (Fig. 4.2) [42]. Urinary tract infections in mice by mannose- or

Table 4.4 Human Blood Group Structures [1,35]

Blood group	Structure
A	GalNAcα1-3Galβ- 2 \| Fucα1
B	Galα1-3Galβ- 2 \| Fucα1
H	Fucα1-2Galβ-
Lewis-a (Le[a])	Galβ1-3GlcNAcβ- 4 \| Fucα1
Lewis-b (Le[b])	Galβ1-3GlcNAcβ 2 4 | | Fucα1 Fucα1
P$_1$	Galα1-4Galβ1-4GlcNAcβ1-3Galβ1-4Glcβ-
P	GalNAcβ1-3Galα1-4Galβ1-4Glcβ-
p[k]	Galα1-4Galβ1-4Glcβ-
p	Galβ1-4Glcβ-

Galα1-4Gal-specific *E. coli* strains were prevented by methyl α-D-mannoside [43] or globoside [44], respectively.

Human influenza viruses contain a glycoprotein (hemagglutinin) that binds to oligosaccharide sequences containing terminal sialic acid. Recently, the three-dimensional structure of the influenza virus hemagglutinin complex with Neu5Acα2-6Galβ1-4Glc (sialyllactose) was determined [45]. The interactions revealed may be instrumental for the design of general antiviral drugs based on receptor analogue inhibitors.

Monoclonal antibodies against tumor-associated carbohydrate antigens are used in the diagnosis of cancer; for example, antibodies directed against the sialyl-Le[a] antigen serve in the diagnosis of gastrointestinal cancer [10]. Immunotherapy of cancer patients with monoclonal antibodies directed toward cancer-associated antigens, or via vaccination with glycoconjugates or neoglycoconjugates, has been investigated [11].

Table 4.5 Proposed Receptor Structures for Pathogens and Toxins

Receptors	Pathogen or toxin
Galα1-4Gal	P-fimbriated *E. coli*
Manα-R	Type 1 fimbriated *E. coli*
Galβ1-4GlcNAc	Staphylococcus saprofyticus
GlcNAcβ1-3Gal	Streptococcus pneumoniae
Neu5Acα2-3Galβ1-4GlcNAc	Mycoplasma pneumoniae
Galβ1-3GalNAcβ1-4(Neu5Acα2-3)Galβ1-4Glc-	Cholera toxin
Galα1-4Galβ1-4Glc	Shiga toxin

Source: Ref. 6.

**Column with macro-
beads, modified with
Galα1-4 Galβ-O**

**Spherical microbead
(0.2-5μm) with enzyme
and Galα1-4 Galβ-O**

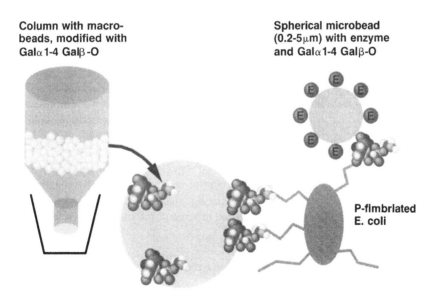

**P-fimbriated
E. coli**

Figure 4.2 An assay of P-fimbriated *E. coli* based on the Galα1-4Gal receptor structure [42].

A specific assay of a glycosyltransferase, N-acetylglucosaminyltransferase V (GnT-V), was recently reported [46]. The enzyme, which initiates GlcNAcβ1-6Man branching and can be elevated two-fold in certain tumors, was detected using the trisaccharide structure GlcNAcβ1-2Manα1-6Manβ as a specific acceptor.

Chemical and Enzymatic Methods for the Production of Glycoconjugate Oligosaccharides

The chemical synthesis of oligosaccharides is considered to be more complex than that of other biopolymers. This is due to the α- or β-configuration between glycosyl units and the similar and relatively low reactivity of the multiple hydroxyl groups of oligosaccharides. Among the methods used to form glycosidic linkages, the classical Koenigs-Knorr method (halogen as leaving group; usually bromo- or chloroactivation; Fig. 4.3) [47], and methods using the trichloroacetimidate and the thioethyl groups as leaving groups, may be mentioned [13,14]. In the former, usually more or less expensive, toxic, and in some cases explosive, heavy metal salts [Ag-triflate, Hg(CN)$_2$, AgClO$_4$] of differing reactivities are used as catalysts. A number of different protection groups (e.g., allyl, acetyl, benzyl, benzoyl, trityl groups) are used to promote regioselective reactions and to adjust the reactivities of the donor electrophile and acceptor hydroxyl group in order to achieve good stereoselectivity and yield.

The selective chemical synthesis of a simple disaccharide requires several reaction steps (Fig. 4.3), and synthesis of trisaccharides often requires 8–15 steps, depending on the complexity of the desired product. The presence of the carboxyl group at the anomeric center of N-acetylneuraminic acid and the adjacent methylene carbon renders the chemical synthesis of the important sialyloligosaccharides (Tables 4.1-4.3) especially difficult (reported yields 20-40%) despite recent progress [48]. The unnatural β-sialoside has to be separated from the desired α-sialoside [49,50].

Thus, it is obvious that enzymes, which are efficient and highly selective catalysts at room temperature and in neutral water, are especially suited for application in the synthesis of complex oligosaccharides. Both glycosidases and glycosyltransferases have been used to promote the stereospecific synthesis of oligosaccharides.

SYNTHESIS WITH GLYCOSIDASES

Introduction

Glycosidases (EC 3.2) are ubiquitous in nature and have a vital catabolic function in living organisms. They catalyze the hydrolysis of simple glycosides,

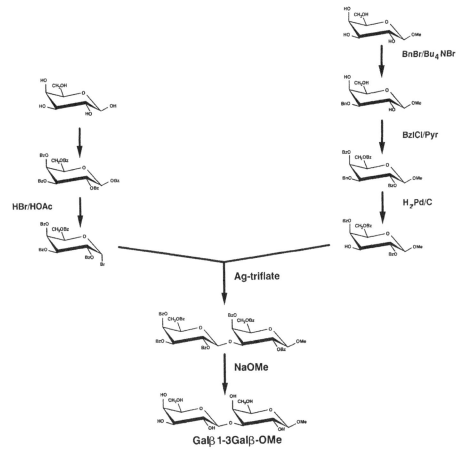

Figure 4.3 Chemical synthesis of methyl 3-*O*-β-D-galactopyranosyl-β-D-galactopyranoside [47].

oligosaccharides, and polysaccharides and are important for the trimming of *N*-linked glycoproteins in vivo. Several inherited metabolic disorders are associated with the lack of specific glycosidases, and malignant transformation of cells has in several cases been shown to be accompanied by changed levels of glycosidases [51].

Historically, glycosidases were among the first enzymes to be investigated, and studies on glycosidases have been important for the formulation of some of the basic principles of enzyme action. Early examples are the stereospecific hydrolysis of α- and β-glucosides with yeast extract, as reported in 1894 [52], and the stereospecific synthesis of various alkyl and aryl β-glucosides with β-glucosidase suspended in almost pure organic solvent, reviewed in 1913 [53]. Glycosidases are used in various applications: in the production of invert sugar, in the hydrolysis of lactose in milk, for visualization in diagnostical reagents, and recently for the structural analysis [54] and synthesis of complex carbohydrates [36,37].

Enzymic Properties
Glycosidases active on O-, N-, S-glycosyl compounds (EC 3.2.1, EC 3.2.2, and EC 3.2.3, respectively) are known [55-57]. β-Galactosidase from *E. coli*, which is classified as an O-glycosidase, shows some activity on S-, N-, and F-glycosides as well [56]. This review concentrates on O-glycosidases, but a few N-glycosidases (e.g., β-aspartyl-N-acetylglucosaminidase, glycopeptide N-glycosidase, see ref. 55) are of relevance to the glycoconjugate field. The S-glycosidases may be useful for the synthesis of thioglycosides.

O-Glycosidases are classified as hydrolases and are named according to the type of bond hydrolyzed (Table 4.6) [55]. However, many glycosidases also catalyze the transfer of glycosyl residues to various hydroxyl group containing compounds other than water, and hydrolysis thus represents a special case. The overall reaction catalyzed can be written:

$$glycosyl{-}OR + R'OH \longrightarrow glycosyl{-}OR' + ROH$$

where glycosyl—OR symbolizes the glycosyl donor substrate and R'OH the acceptor. The "glycosyl" moiety (mono- or oligosaccharide group) is often referred to as the glycon part and the R group as the aglycon part. Glycosidases catalyze the synthesis of glycosidic linkages via reversed hydrolysis (equilibrium approach; R = H; R' = organic group including a sugar group) or via transglycosylation (kinetic approach; R and R' = organic group, including a sugar group).

Mechanism of Action
Even though the first enzyme whose three-dimensional structure was elucidated was a glycosidase—lysozyme in 1967 [58]—the catalytic mechanism of most glycosidases has not been elucidated in any detail. To facilitate the synthesis of efficient inhibitors, the structure of two subtypes of type A influenza virus neuraminidases have been determined to medium resolution [59]. Site-specific mutagenesis experiments in glycosidases have been reported [186, 187] to define the role of active site residues.

Table 4.6 Examples of Glycosidases [51,54,55]

Glycosidases	Linkage cleaved
exo-	
α-L-Fucosidase	Fucα-OR
α-D-Galactosidase	Galα-OR
β-D-Galactosidase	Galβ-OR
α-D-Mannosidase	Manα-OR
β-D-Mannosidase	Manβ-OR
α-*N*-Acetylgalactosaminidase	GalNAcα-OR
β-*N*-Acetylgalactosaminidase	GalNAcβ-OR
β-*N*-Acetylglucosaminidase	GlcNAcβ-OR
β-Aspartyl-*N*-acetylglucosaminidase	GlcNAcβ1-Asn
endo-	$\downarrow$
endo-β-Galactosidase	... GlcNAcβ1-3Galβ1-4(3)GlcNAc(or Glc ...
	$\downarrow$
Glycopeptide *N*-glycosidase	... Manβ1-4GlcNAcβ1-4GlcNAcβ1-Asn ...
	$\downarrow$
endo-α-*N*-Acetylgalactosaminidase	Galβ1-3GalNAcα1-Ser/Thr ...
	$\downarrow$
endo-Glycoceraminidase[a]	... Galβ1-4Glcβ1-1Cer

Source: M. Ito and T. Yamagata, *J. Biol. Chem. 261*, 14278–14282 (1986).

Analogous mechanisms which involve general acid-base catalysis have been proposed for hen's egg white lysozyme, influenza virus neuraminidase, *E. coli* β-galactosidase, and other glycosidases (Fig. 4.4) [51,56, 59–62,186,187]. The glycosyl–oxygen bond is cleaved (as shown by $H_{18}O$ experiments). In the case of β-galactosidase, this is further supported by the fact that cleavage of di- or trinitrophenyl β-thiogalactosides gives galactose and the corresponding thiophenols [56].

It has been proposed [61,62] that in the case of hen's egg white lysozyme the side chain carboxylic group of Glu 35 acts as a general acid that catalyzes the cleavage of the glycosyl-oxygen bond, and the oxocarbonium ion thus formed (in a "half-chair" conformation) is electrostatically stabilized by the ionized form of the side chain carboxylic group of Asp 52, which enables the aglycon to diffuse away. For the final catalytic step, it is proposed that water or another glycosyl acceptor is deprotonated by Glu 35, which increases the

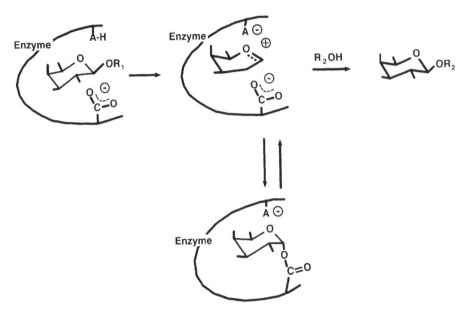

Figure 4.4 Proposed schematic mechanism for glycosidase–catalyzed reactions.

nucleophilicity of the acceptor and thus facilitates the formation of the new glycosidic bond. The hydrophobic environment and the hydrogen bonding increase the pK_a values of Glu 35 and Asp 52 by about one unit.

A covalent enzyme-glycosyl ester intermediate has been proposed for some other glycosidases, E. coli (Lac Z), β-galactosidase [60]. Both the latter type of intermediate and the ion-pair glycosyl–enzyme intermediate can explain the stereospecific reactions of glycosidases.

Substrate Specificity

Glycosidases are generally classified as endoglycosidases that act on glycosidic linkages within oligo- or polysaccharides and exoglycosidases that are active on terminal linkages and release monosaccharides (Table 4.6).

For synthetic purposes it is of interest to know the range of substrates that can be used in synthesis. In general, glycosidases are highly specific for the glycon part of the substrate, with a wider tolerance for structural variations of the aglycone moiety. The main features of the glycon part required for activity are ring form (e.g., D-galactopyranosides for D-galactosidases and L-fucopyranosides for L-fucosidases), correct anomeric configuration, and correct orientation of hydroxyl groups [51,56,63,64].

The requirements for the ring structure and the anomeric configuration are absolute, but some tolerance to the orientation of hydroxyl groups and to substituents on the ring atoms has been observed. Thus, p-nitrophenyl β-glycosides of both N-acetyl-D-glucosaminide and N-acetyl-D-galactosaminide can be used as glycosyl donors in reactions catalyzed by jack bean β-N-acetyl-D-hexosaminidase (V_{max} = 137 and 71 U/mg, respectively; K_m = 0.64 and 0.31 mM, respectively) [65]. The β-mannosidase from *Aspergillus niger* shows some activity to β-glucosides and β-galactosides [66].

Substitution of C-6 of hexose glycones is tolerated in some cases but is accompanied by changed values of K_m and V_{max}. An example is an α-D-galactosidase from *Vicia faba*, which in addition to p-nitrophenyl α-D-galactoside (V_{max} = 25.5 U/mg, K_m = 0.38 mM), also hydrolyzes p-nitrophenyl α-D-fucoside (i.e., substitution of the CH_2OH in galactose to CH_3) (V_{max} = 24 U/mg, K_m = 4.8 mM) and p-nitrophenyl α-L-arabinoside (i.e., substitution of $-CH_2OH$ to $-H$) (V_{max} = 16.4 U/mg, K_m = 14.3 mM) [64].

The effects of replacing the glycosidic oxygen of glycosides have not been extensively investigated. Alkyl or phenyl β-D-thiogalactosides are not substrates for the β-D-galactosidase from *E. coli* but can be used as specific inhibitors of the enzyme [56]. Aromatic β-D-thiogalactosides substituted with electron-withdrawing groups are substrates, and the rate of hydrolysis of 2,4,6-trinitrophenyl β-D-thiogalactopyranoside is comparable to that of Galβ-OPhNO$_2$-o [56]. Moreover, the *E. coli* enzyme cleaves C-F and C-N glycosidic linkages. The β-D-galactosyl azide is a good substrate (k_{cat} = 26/s, K_m = 2.8 mM) [67].

Aglycon Specificity

A wide variety of aglycon groups are tolerated by most glycosidases, although the kinetic parameters are influenced by the aglycon structure. Aromatic glycosides (e.g., nitrophenyl glycosides) are often good substrates with a higher V_{max} than alkyl glycosides or saccharides (i.e., the agylcon = sugar). However, an α-fucosidase from *Aspergillus niger* is active on Fucα1-6GlcNAc linkages in various glycoconjugate glycans but not on p-nitrophenyl α-L-fucopyranoside [68]. A progressive decrease in K_m for increasing chain length of alkyl glycosides has been observed with several glycosidases, which indicates a hydrophobic character of the aglycon site of these enzymes [57].

Another aspect of glycosidase aglycon specificity is the selectivity toward positional isomers (regioselectivity) and the "branch" specificity. The regioselectivity is often not pronounced, but in some cases a high selectivity toward certain linkages may be obtained (Table 4.7). Thus, the *E. coli* β-galactosidase is highly active on Galβ1-6GlcNAc but shows a very low activity toward the (β1-3)-linked isomer [69], jack bean α-mannosidase

Table 4.7   Relative Rate of Hydrolysis of β-Linked Gal-GlcNAc Isomers [69,70]

	Enzyme source		
Substrate	Bovine testes	*Diplococcus pneumoniae*	*E. coli*
Galβ1-3GlcNAcβ-OR	100	0	0.2
Galβ1-4GlcNAcβ-OR	80	100	45
Galβ1-6GlcNAcβ-OR	4	4	100

cleaves Manα1-2Man and Manα1-6Man about 15 times as fast as the (α1-3)-linked isomer [71], and α-mannosidase I from *Aspergillus saitoi* specifically cleaves Manα1-2Man but not (α1-3) or (α1-6)-linked dimannose [72]. The source of the enzyme is important, and in contrast to the *E. coli* enzyme, the β-galactosidase from *Diplococcus pneumoniae* hydrolyzes Galβ1-4GlcNAc-OR with high selectivity over the (β1-6)-linked isomer [69].

Longer aglycon structures may be recognized. Examples of this are the cleavage of the α1-2 linkage of Fucα1-2Galβ1-4Glc but the inactivity toward Fucα1-2Galβ1-3(Fucα1-4)GlcNAc shown by α-fucosidase from human serum [73] and the branch specificity of *E. coli* β-galactosidase toward biantennary lactosamine glycans [74]. Another example is neuraminidase from human influenza A$_2$ virus, which cleaves sialic acid from Neu5Acα2-6Galβ1-4GlcNAc but has low activity toward the same oligosaccharide linked to asparagine [75]. Endoglycosidases often show high sequence selectivity for the nonreducing end of saccharides [51,54].

Occurrence and Isolation

The practical applications of glycosidases are facilitated by the high availability of these enzymes with different properties in microorganisms, plants, and animal cells. Isolation by such conventional techniques as extraction, fractionation by salts, and treatment with heat, low/high pH, organic solvent, and (affinity) chromatography, is usually straightforward [51,63,76–79]. Several glycosidases are glycoproteins, and lectins have been used as affinity ligands [76], in addition to monosaccharides [77], or sugar analogues [78], attached to agarose via *N*- or *S*-glycosidically linked spacers [79]. An aqueous two-phase system was used for large-scale preparation of *E. coli* β-galactosidase [80].

While pure glycosidases are required for structural studies on glycoconjugates, crude enzyme preparations can be successfully used in the synthesis of glycosidic bonds by transglycosylation. In the case of bacteria or yeast, the enzyme may be used in situ. If, for example, β-galactosidase is used in the synthesis of β-linked Gal-GlcNAc by means of a transglycosylation reaction (e.g., with lactose and GlcNAc as donor and acceptor substrates, respectively), contaminating glycosidases other than β-galactosidases are of less importance because transglycosylation of lactose is normally considerably faster than reversed hydrolysis. Several examples of the successful synthesis of oligosaccharides with crude glycosidase preparations can be found in the literature [81-83].

Conversely, synthesis by reversed hydrolysis of the same type of structure from galactose and N-acetylglucosamine requires pure β-galactosidase to minimize the formation of GlcNAc-Gal or of α-linked Gal-GlcNAc with contaminating glycosidases.

The stereospecificity, glycon specificity, sequence specificity, simple reaction systems, and easy availability of different glycosidases are all properties that have been of great value in the analysis of glycoconjugate glycan structures and for the selective synthesis of these structures.

Synthesis of Glycoconjugate Glycan Sequences with Glycosidases

The three types of glycosidase reactions—hydrolysis, transglycosylation, and reversed hydrolysis—have all been used for the preparation of oligosaccharides. Synthesis via transglycosylation has been predominantly used. Glycosidases have been mainly used for synthesis of shorter carbohydrate sequences of glycoconjugates (di- and trisaccharides), but as the methods for synthesis with exoglycosidases are more developed and endoglycosidases increasingly applied, the synthesis of higher structures can be expected.

Hydrolysis

Chemical synthesis of Galα1-4Gal (the receptor structure for the common uropathogen P-fimbriated *E. coli*) was facilitated by the preparation of large quantities of digalacturonic acid by the controlled pectinase-catalyzed hydrolysis of pectin [84]. The disaccharide receptor structure and glycosides, suitable as inhibitors or for the attachment to proteins or chromatography material, were conveniently obtained via chemical modification of the digalacturonic acid.

Oligosaccharide chains have been selectively released from glycolipids and glycoproteins using endoglycosidases [51,54] (cf. Table 4.6). An example is the *Diplococcus pneumoniae endo-β*-galactosidase, which selectively releases blood group A and B trisaccharides from the nonreducing end of glycoconjugates [85]. A number of endoglycosidases are commercially available. Recombinant *N*-glycanase (EC 3.5.1.52), which releases asparagine-linked carbohydrate chains in their intact form, is also available.

In a third type of approach, a crude oligosaccharide mixture is produced in a glycosidase-catalyzed transglycosylation reaction, and another glycosidase with a different regioselectivity is then used in a second reaction for the selective hydrolysis of contaminating oligosaccharide isomers. This method has been used to facilitate purification of Galβ1-3GlcNAc and Galβ1-3GalNAc synthesized from lactose and GlcNAc or GalNAc, respectively, using bovine testes and *E. coli* β-galactosidase in the transglycosylation and hydrolytic reactions, respectively [86]. The *E. coli* enzyme hydrolyzes the (β1-4) and (β1-6)-linked isomers with high selectivity (Table 4.7).

Synthesis by Transglycosylation

In the synthesis of glycoconjugate glycan sequences by transglycosylation, a glycoside (e.g., an oligosaccharide or an alkyl or an aryl glycoside) is used as glycosyl donor and the glycon part of the glycoside is transferred to a suitable hydroxyl group containing acceptor. The reaction is stereospecific and occurs with retention of the donor anomeric configuration. Rabaté reported the synthesis of various β-glucosides by transglycosylation reactions with β-glucosidases 1935 [87]. The preparative syntheses of the glycoconjugates in common disaccharides Galβ1-4GlcNAc and Galβ1-3GlcNAc employing *Lactobacillus bifidus* or mammalian β-galactosidases were reported in 1955 [82] and in 1956 [81], respectively. Several glycoconjugate oligosaccharide structures have now been synthesized with glycosidase-catalyzed transglycosylation reactions [36,37,81-83,86,88-92].

Compared with reversed hydrolysis, higher yields may be obtained with considerably shorter reaction time, because of the higher reactivity of the donor glycoside. Furthermore, as mentioned above, crude enzyme preparations may be used, whereas the equilibrium approach requires pure glycosidase preparations. However, in the equilibrium approach substrates can be reused while the donor glycoside is consumed in the transglycosylation reaction (i.e., forming hydrolysis and transglycosylation products).

A number of glycosides suitable as glycosyl donors in transglycosylation reactions are available commercially, or are easy to prepare enzymatically [90] or by means of standard chemical procedures [13,14]. Cheap oligosaccharides (lactose, raffinose, oligomannosides) have been used successfully as donors [36,37,88,90]. Although more expensive than the oligosaccharides, nitrophenyl glycosides are often used as donors because of their high reactivity and the possibility of estimating the amount of donor consumed during the reaction by measuring the amount of liberated nitrophenol spectrophotometrically [88]. As indicated above, the source of the enzyme is important when choosing a suitable donor; for example, o-nitrophenyl β-galactoside is an excellent substrate for β-galactosidase from *E. coli*, whereas calf intestine β-galactosidase hydrolyzes lactose more rapidly then the nitrophenyl glycoside [93].

The principles of transglycosylation can be further illustrated by Scheme 1, proposed for the *E. coli* β-galactosidase [56,60]:

$$E \cdot Gal \cdot H_2O \rightleftharpoons Gal\text{-}OH$$

$$\updownarrow H_2O$$

$$\underset{\text{-ROH}}{Gal\beta\text{-}OR + EH \rightleftharpoons EH \cdot Gal\beta\text{-}OR \rightleftharpoons E - Gal}$$

$$\updownarrow HOA$$

Scheme 1 $\qquad E \cdot Gal \cdot HOA \rightleftharpoons Gal\beta\text{-}OA$

where $Gal\beta$-OR is the donor glycoside (e.g., lactose or $Gal\beta$-OPhNO$_2$-o), EH is the enzyme, HOA is the acceptor, and $Gal\beta$-OA is the transglycosylation product.

In the initial phase of the reaction, the glycosyl-enzyme intermediate will be preferentially formed from the donor glycoside. The intermediate can react with a wide range of hydroxyl group containing acceptors other than water, including monosaccharides, oligosaccharides, and their analogues, as well as aliphatic or aromatic alcohols, and a number of different compounds can thus prepared from the donor glycoside. The rate of transglycosylation relative to hydrolysis is dependent on the structure of the acceptor and this selectivity is different from enzyme to enzyme.

It should be observed that the glycosyl donor can also act as acceptor, and this property has been used for the facile synthesis of various disaccharides such as Galα1-3Galα-OMe, Galα1-3Galα-OPhNO$_2$-p, Manα1-2Manα-OPhNO$_2$-p [88], and Manα1-6ManαOMe [91] starting from GalαOMe, Galα-OPhNO$_2$-p, and Manα-OMe, respectively. Thus, if this reaction is not desired, a high ratio of glycosyl acceptor to glycosyl donor should be used. *E. coli* β-galactosidase has been shown to catalyze so-called direct transglycosylation of lactose, in which the aglycon (glucose) does not leave the acceptor binding site but changes position, and a lactose isomer, Galβ1-6Glc (allolactose), is formed [94].

The product glycoside can, of course, also react with the glycosyl enzyme intermediate to form higher oligosaccharide products. This capability has been used for the synthesis of preparative amounts of trisaccharide structures from monosaccharide glycosides. Thus, Manα1-2Manα1-2Manα-OMe was pre-

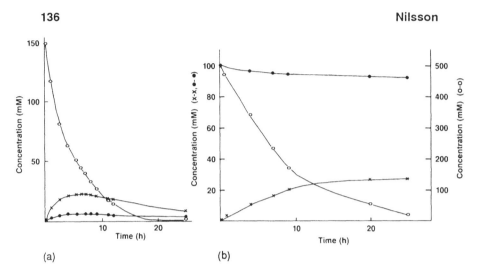

Figure 4.5 Formation of (a) Galα1-2Galα-OPhNO₂-*p* (•) and Galα1-3Galα-OPhNO₂-*p* (x), and (b) Galα1-3Galα-OEtBr (x), from Galα-OPhNO₂-*p* (o) and Galα-OEtBr (•). (From ref. 91.)

pared from Manα-OPhNO₂-*p* and Manα-OMe employing jack bean α-mannosidase [88].

The glycosidase–catalyzed transglycosylation reactions show similarities with other kinetic reactions, such as protease-catalyzed transamination [95]. The rapid accumulation of product is followed by a second phase in which the equilibrium concentrations of reactants and products are slowly achieved [91]. Initially, the rate of product formation will depend on the partition ratio of the glycosyl enzyme intermediate between the transfer and hydrolytic reactions (see Scheme 1). As the donor glycoside is consumed and the transglycosylation product accumulates, the hydrolysis rate of the product glycoside eventually becomes larger than the rate of synthesis. Secondary hydrolysis will then decrease the yield of product glycoside to that of an equilibrium-controlled process.

Therefore, in general, a good maximum yield is favored not only by the use of a high concentration of an efficient acceptor, which gives a high transfer/hydrolysis ratio, but also by the use of a highly reactive donor substrate having a high k_{cat}/K_m ratio compared with that of the product glycoside. The reaction should be followed (e.g., by HPLC) and stopped at the point of maximum yield. Two examples of product formation profiles are shown in Figure 4.5. In Figure 4.5a *p*-nitrophenyl α-D-galactopyranoside is used as both donor and acceptor, and in Figure 4.5b a high concentration of a separate acceptor glycoside is used [91]. In the latter case the yield increases until practically all Galα-OPhNO$_2$-p is consumed.

Yield and Regioselectivity. The demands on the catalyst in oligosaccharide synthesis are higher than in, for instance, peptide synthesis. Both types of reaction involve condensation between a donor electrophile and an acceptor nucleophile; but hydroxyl groups are less nucleophilic than amino groups and, in addition to the possibility of alternative configuration (α- or β-), the multiple hydroxyl groups of the acceptor sugar put high demands on acceptor specificity.

As mentioned above, glycosidases catalyze stereospecific reactions and are highly specific for the glycon part of the substrate, but show a less pronounced selectivity for the aglycon part of the glycoside substrate or for the acceptor in a transglycosylation reaction. In other words, a wide range of structures can be synthesized from a given glycosyl donor with a single glycosidase, and the yield and the regioselectivity in transglycosylation reactions tend to be relatively low.

Several positional isomers of the product oligosaccharide are often obtained, and if an acceptor with a free reducing end (i.e., with a hydroxyl group at C-1) is used, the products can adopt α- or β-configuration at the reducing end. This leads to complex product mixtures.It is thus obvious that isolation of the desired product may be a considerable problem, which lowers the isolated yield.

Furthermore, glycosidases often catalyze the predominant formation of undesired (1-6)-linkages with hexoses as glycosyl acceptors, which in part may be attributed to the higher nucleophilicity of primary hydroxyl groups [56,57]. For example, most α-galactosidases tested catalyze transfer to primary hydroxyl groups [64], and β-D-galactosidase from *E. coli* predominantly synthesize (β1-6)-linked Gal-Gal, Gal-Glc, Gal-GalNAc, and Gal-GlcNAc with lactose or Galβ-OPhNO$_2$-p as glycosyl donors and with galactose, glucose [93,94], *N*-acetylgalactosamine [96], and *N*-acetylglucosamine [97], respectively, as acceptors.

Influence of Enzyme Source and Reaction Time. The source of the enzyme is important (Table 4.8). The regioselectivity of these reactions was good (up to 90%), except the bovine testes β-galactosidase-catalyzed synthesis of (β-3)- and (β1-4)-linked Gal-GlcNAc. We have used some of these reactions for preparation of disaccharides on a kilogram scale and with yields exceeding 20%.

The rate of hydrolysis of isomeric oligosaccharides can be useful for predicting the regioselectivity of transglycosylation. Thus, in the case of *E. coli* β-galactosidase, the relative rates of hydrolysis of Gal-Glc were β1-6 >

Table 4.8 Synthesis of Disaccharides with Glycosidases from Different Sources

Enzyme	Source	Structure synthesized	Ref.
β-Galactosidase	*E. coli*	Galβ1-6GlcNAc	97
	Mammary gland	Galβ1-3GlcNAc	81
	Bovine testes	Galβ1-3GlcNAc	81
		Galβ1-4GlcNAc	
	Lactobacillus bifidus	Galβ1-4GlcNAc	82
β-*N*-Acetylgalactosaminidase	Bovine testes	GalNAcβ1-6GalβOMe	[a]
	Chamelea gallina	GalNAcβ1-3GalβOMe	92
β-*N*-Acetylglycosaminidase	Jack bean	GlcNAcβ1-6GalβOMe	83
	Chamelea gallina	GlcNAcβ1-3GalβOMe	92

[a]K. G. I. Nilsson, unpublished.

β1-4 > β1-3, and the synthetic rates decreased in the same order [93]. With the β-galactosidases from calf intestine and bovine testes, the reversed order for both hydrolysis and transglycosylation was observed [93]. This is not a rule and, for example, α-L-fucosidase from *Chamelea gallina* rapidly hydrolyzes Fucα1-2Gal [98], but the transglycosylation reactions with either Galα-OMe or Galβ-OMe as acceptors give predominantly the (α1-6)-linked fucosyl-galactosides [92].

The rate of hydrolysis may have more direct effects on the regioselectivity of transglycosylations. For example, a more rapidly formed product isomer may be more susceptible to secondary hydrolysis than a more slowly formed product, and the ratio of product isomers will depend on when the reaction is terminated [91]. Thus, in the initial phase of the synthesis of Galα1-3Galα-OPhNO2-p from Galα-OPhNO2-p, the regioselectivity was about 85%, whereas the (α1-3)-linked digalactoside constituted only about 65% of the disaccharide glycosides formed when practically all the donor was consumed (Figure 4.5a). Provided an excess of glycosyl donor is used and the difference in secondary hydrolysis between isomers is high, the same enzyme can be used for the preferential synthesis of different product isomers (unpublished results). Furthermore, one glycosidase can be used for synthesis and another glycosidase with a different linkage specificity for hydrolysis of unwanted isomers (see "Hydrolysis," p. 133).

Influence of Acceptor Structure on Regioselectivity. The regioselectivity of glycosidase-catalyzed transglycosylations can be changed by using different α- or β-glycosides as acceptors (Table 4.9) [88,92]. Both the anomeric configuration and the structure of the aglycon of the acceptor are important. The effect of changing the anomeric configuration is shown in Table 4.9 for α-galactosidase, β-galactosidase, and β-hexosaminidase. With α-galactosidase, methyl β-D-galactopyranoside gave mainly the $(\alpha$-6)-linked digalactoside, whereas the α-anomer gave almost exclusively the (1-3)-linked disaccharide. The reverse result was found with the β-glycosidases. The ratios between formed product isomers (1-3:1-6) were drastically changed—by factors of about 30 and 100 for the α- and β-galactosidase, respectively. As expected, the effect is different for glycosidases from different sources. Thus, while the β-hexosaminidase from *Chamelea gallina* produces (β1-3)-linked Gal-GlcNAcβ-OMe, the jack bean enzyme give the (β1-6)-linked disaccharide glycoside [92].

The nature of the acceptor aglycon may also have a pronounced influence on the regioselectivity (Table 4.9). Thus, with α-galactosidase, the (α1-3)-linked digalactoside was almost exclusively formed when *p*-nitrophenyl α-D-galactopyranoside was the acceptor, whereas with the corresponding *o*-nitrophenyl glycoside the (1-2)-linked product preponderated [88].

The size or hydrophobicity of the acceptor aglycon also seems to be important. Thus, with bovine testes β-galactosidase, (β1-3)- and (β1-4)-linked

Table 4.9 Influence of the Acceptor Glycoside Structure on the Regioselectivity of Transglycosylations [88,92]

Enzyme	Acceptor	Main glycoside formed
α-Galactosidase (coffee bean)	Galα-OMe	Galα1-3Galα-OMe
	Galβ-OMe	Galα1-6Galβ-OMe
	Galα-OPhNO$_2$-o	Galα1-2Galα-OPhNO$_2$-o
	Galα-OPhNO$_2$-p	Galα1-3Galα-OPhNO$_2$-p
β-Galactosidase (*E. coli*)	Galα—OMe	Galβ1-6Galα-OMe
	Galβ-OMe	Galβ1-3Galα-OMe
	GlcNAcβ-OEtSiMe$_3$	Galβ1-3GlcNAcβ-OEtSiMe$_3$
	GlcNAcβ-OMe	Galβ1-3GlcNAcβ-OMe and Galβ1-4GlcNAcβ-OMe
β-N-Acetylglycosaminidase (*Chamelea gallina*)	Galα-OMe	GlcNAcβ1-6Galα-OMe
	Galβ-OMe	GlcNAcβ1-3Galβ-OMe

Gal-GlcNAc-OMe are formed in about equal amounts when GlcNAcβ-OMe is used as acceptor, while (β1-3)-linked Gal-GlcNAcβOEtSiMe3 is formed with high regioselectivity ($\approx$90%) when the trimethylsilylethyl glycoside is the acceptor [83]. Similarly, with α-mannosidase from jack bean, the ratio of (α1-2)- and (α1-6)-linked products was 5:1 with methyl α-D-mannopyranoside as acceptor, whereas the ratio was 19:1 with the corresponding p-nitrophenyl glycoside [88].

The effects above reflect the properties of the glycosyl acceptor binding sites thought to exist in glycosidases [57,94,99]. The existence of hydrophobic regions in the acceptor site has been proposed for several glycosidases [57,95]. Thus, the interaction of the enzyme and the acceptor will be influenced by the nature of the aglycon and the anomeric configuration. Furthermore, steric interactions between the acceptor aglycon and the enzyme may be important.

The foregoing strategy of controlling the regioselectivity thus allows the use of one glycosidase for the preferential synthesis of several different linkages. Glycosidases known to prefer primary hydroxyl groups (i.e., formation of 1-6 linkages), such as the β-galactosidase from *E. coli*, can now be used for preparative synthesis of other linkages, thus extending the number of structures that can be synthesized with easily available glycosidases.

Isolation of Products and Yields

Purification of products is simplified when acceptor glycosides are used, since no anomerization of the product glycosides occurs. Thus, although the reactions are not completely regioselective, column chromatography (Sephadex, Biogel, silica) of the isomeric products are usually straightforward, and the purity of the products exceeded 99%.

The interaction between the aglycon and the separation media may be utilized. Thus, the α-linked nitrophenyl digalactoside isomers obtained with α-galactosidase, were separated in one chromatographic step using Sephadex G10 (Fig. 4.6) or preparative HPLC with C18-silica (unpublished result).

The method has been used for the facile synthesis of disaccharide glycosides on a gram-kilogram scale and with molar yields of 10-50% calculated on the limiting donor substrate. Because of the simplicity of these reactions, the yields are high enough to make the method attractive. Recovery of excess of reagents by column chromatography or precipitation is usually straightforward. Table 4.10 summarizes glycoconjugate disaccharide structures synthesized with the procedure.

Synthesis of Trisaccharides

The examples above relate to synthesis of disaccharide structures present in glycoconjugates. The synthesis of some trisaccharides has been achieved as well. As mentioned above, Manα1-2Manα1-2Manα-OMe is formed in a

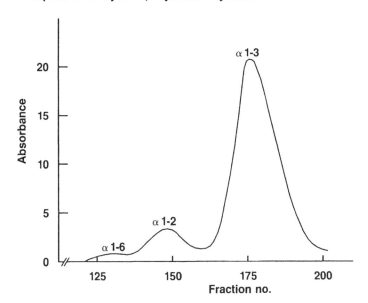

Figure 4.6 Separation of Sephadex G10 of α-linked p-nitrophenyl digalactosides formed in coffee bean α-galactosidase catalyzed transglycosylation of Galα-OPhNO$_2$-p [88].

"one-pot" reaction from Manα-OPhNO$_2$-p and Manα-OMe with jack bean α-mannosidase as catalyst. Similarly, with bovine testes β-galactosidase, Galβ1-3Galβ1-4GlcNAcβ-OEtSiMe$_3$ is formed from Galβ-OPhNO$_2$-o (glycosyl donor) and GlcNAcβ-OEtSiMe$_3$ (acceptor) [83]. Interestingly, the β-galactosidase seems to be selective for the acceptor disaccharide Galβ1-4GlcNAcβ-OEtSiMe$_3$, since the β1-3 isomer, which was the major disaccharide glycoside formed in the reaction, was a poor acceptor. Galα1-3Galβ1-4GlcNAcβ-OEt, a cancer-associated structure [100], was obtained by the sequential use of β-galactosidase and α-galactosidase (coffee bean) [101]. The β-galactosidase gave Galβ1-4GlcNAcβ-OEt, which was used as acceptor in the second reaction with Galα-OPhNO$_2$-p as donor and α-galactosidase as catalyst.

The yields and the regioselectivities of the foregoing reactions are generally low. The bovine testes β-galactosidase reaction, however, was quantitative (~50%), and this enzyme seems to be useful for the preparative synthesis of various trisaccharides and higher oligosaccharide structures [83].

The use of exoglycosidases for synthesis of shorter fragments (e.g., disaccharides) and the use of endoglycosidases for condensation of these fragments

Table 4.10 Some Glycosides of Glycoconjugate Disaccharide Structures
Synthesized with Glycosidases

Disaccharide structure	Ref.
α-N-Acetylgalactosaminidase	
GalNAcα1-3Galα-OMe	92
β-N-Acetylgalactosaminidase	
GalNAcβ1-3Galβ-OMe	92
β-N-Acetylglucosaminidase	
GlcNAcβ1-6Galβ-OMe	83
GlcNAcβ1-6Galα-OMe	92
GlcNAcβ1-3Galβ-OMe	92
GlcNAcβ1-6Manα-OMe	83, 92
α-L-Fucosidase	
Fucα1-6Galβ-OMe	92
α-Galactosidase	
Galα1-3Galα-OPhNO$_2$-p	88
Galα1-3Galα-OMe	88
Galα1-3GalNAcα-OEt	83
β-Galactosidase	
Galβ1-3Galβ-OMe	88
Galβ1-3GalNAcα-OEt	83
Galβ1-3GlcNAcβ-OMe	83
Galβ1-3GlcNAcβ-OEtSiMe$_3$	83
Galβ1-4GlcNAcβ-OMe	83
α-Mannosidase	
Manα1-2Manα-OMe	88
Manα1-6Manα-OMe	91

may be a possible strategy for synthesis of higher oligosaccharides with
glycosidases. *endo-β-N*-Acetylglucosaminidase F has transfer activity [102],
and *endo-α-N*-acetylgalactosaminidase from *Diplococcus pneumoniae* was
recently reported to catalyze the formation of glycosides of the type
Galβ1-3GalNAcα-OR [103].

Applications of Synthesized Glycosides

The use of acceptor glycosides provides an approach to glycosides suitable for
various applications. For example, allyl-, benzyl-, 2-bromoethyl-, nitro-

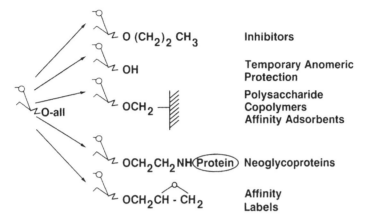

Figure 4.7   Some conversions/applications of allyl glycosides.

phenyl–, thioethyl–, and trimethylsilylethyl glycosides have been conveniently prepared [83,88–92]. The use of other aglycons such as amino acids or peptide derivatives is also possible [104].

Allyl– [105], benzyl– [106], and trimethylsilylethyl glycosides [107] are commonly used for temporary anomeric protection in the chemical synthesis of oligosaccharides, and thioethyl glycosides are used as building blocks in the formation of oligosaccharides [108]. Allyl–, nitrophenyl– [40], and 2-bromoethyloligosaccharide glycosides [41] can after chemical modification be used for the synthesis of glycosides suitable for inhibition studies, for affinity labeling, or for the preparation of neoglycolipids, neoglycoproteins, or affinity adsorbents (Fig. 4.7).

An obvious extension of the method above is to modify one or more of the remaining hydroxyl groups of the acceptor (other than the reducing end) to withdraw these groups from reaction. This approach has not been extensively investigated. The use of methyl 2,3-*O*-isopropylidene-α-D-manno-pyranoside as glycosyl acceptor leads to the formation of Manα1-6Man-(2,3-*O*-isopropylidene)α-OMe with Manα-OPhNO2-*p* as donor and jack bean α-mannosidase as catalyst [36]. This type of product is useful for further chemical synthesis. To this end, the combined use of lipase and glycosidase may be valuable. Selective protection/deprotection of sugar hydroxyl groups by lipase reactions has been reported [109,110].

The foregoing acceptor glycosides can be prepared enzymatically (with glycosidases), and some disaccharide glycosides have been prepared in "one-pot" reactions from a suitable donor glycoside and alcohol (Table 4.11) [90].

In Scheme 2, a simplified scheme for the β-galactosidase-catalyzed reactions, A symbolizes the alcohol and E the enzyme.

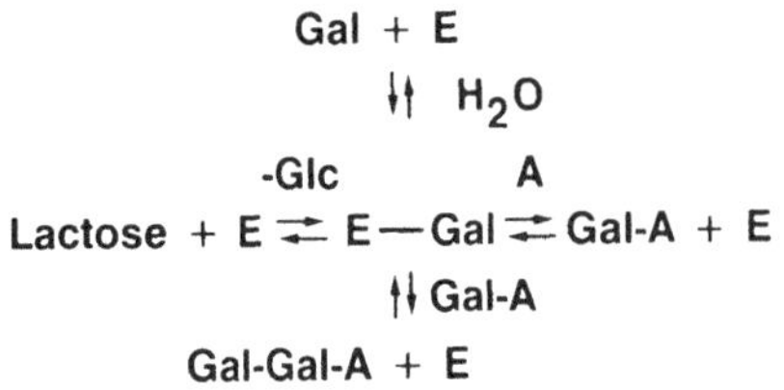

$$
\begin{array}{c}
\text{Gal} + \text{E} \\
\updownarrow\; \text{H}_2\text{O} \\
\text{-Glc}\qquad\text{A} \\
\text{Lactose} + \text{E} \rightleftharpoons \text{E}-\text{Gal} \rightleftharpoons \text{Gal-A} + \text{E} \\
\updownarrow\; \text{Gal-A} \\
\text{Gal-Gal-A} + \text{E}
\end{array}
$$

Although not optimized, about 20 g of allyl β-galactoside and 1 g of Galβ1-3Galβ-OCH$_2$CH$=$CH$_2$ were prepared from lactose and allyl alcohol, employing a few milligrams of β-galactosidase (2.5 mg) [90]. Benzyl alcohol was a much more efficient acceptor than water, since the initial yields of transfer (benzyl β-galactopyranoside) and hydrolysis products (galactose) were practically the same despite the low concentration of benzyl alcohol used (2.5 v/v %).

The enzymatic synthesis of glycopeptides by the combined use of proteases and glycosidases may be an interesting alternative to chemical methods.

Influence of Various Parameters on the Yield

The reported yields of glycosidase-catalyzed transglycosylation have generally been low (10-50%). Anomerization of the products will occur when an acceptor with a free reducing end is used, and chromatography of the products may be a complex task. Thus, even if a good yield is obtained in the reaction, the isolated yield may be low, due to purification problems. In general, as mentioned above, the use of α- or β-glycosides as acceptors is advantageous and

Table 4.11 Synthesis of Mono- and Disaccharide Glycosides from Lactose and Various Alcohols with *E. coli*

Donor	Acceptor	Main glycoside formed
Lactose	Allyl alcohol	Galβ1-OCH$_2$CH=CH$_2$ Galβ1-3Galβ-OCH$_2$CH=CH$_2$
Lactose	Benzyl alcohol	Galβ1-OCH$_2$Ph Galβ1-3Galβ-OCH$_2$Ph
Lactose	Trimethylsilyl ethanol	Galβ1-OEtSiMe$_3$ Galβ1-3Galβ-OEtSiMe$_3$

Source: Ref. 90.

Table 4.12 Effect of *N,N*-Dimethylformamide on α-Galactosidase-Catalyzed Digalactoside Formation

Concentration of cosolvent (%)	Yield (%)	
	Galα1-3Galα-OPhNO$_2$-*p*	Galα1-3Galα-OMe
0	32	32
10		27
15	18	
20		24
30	14	21
45	10	

Source: Ref. 95.

the interaction between the aglycon group of the product glycoside and the adsorbent may facilitate purification (Fig. 4.6).

Organic Solvents. As mentioned above, there are analogies between the kinetic reactions catalyzed by proteases and by glycosidases. In both cases water can react with the enzyme intermediate and lower the yield. One way of minimizing this reaction is to decrease the water concentration by the addition of organic solvent. This approach, however, has not been extensively investigated.

In one study with lysozyme as catalyst, (GlcNAc)5 as donor, and GlcNAcβ-OPhNO$_2$-*p* as acceptor, the yield of (GlcNAc)4GlcNAcβ-OPhNO$_2$-*p* was found to increase upon addition of organic cosolvents [111]. Maximum yields were obtained in 60% aqueous methanol.

In another study with coffee bean α-galactosidase, the yield of Galα1-3Galα-OPhNO$_2$-*p* was found to decrease from 32 to 10% when the concentration of *N,N*-dimethylformamide was increased from 0 to 45% (Table 4.12) [95]. A similar but less pronounced effect was observed for the formation of Galα1- 3Galα-OMe. The decreased yield may be due to a decreased hydrophobic interaction between the enzyme and the acceptor aglycon. The organic solvent did not change the regioselectivity.

Several glycosidases show activity in high concentrations of organic solvents. High yield synthesis of alkyl and aryl β-linked monosaccharide glucosides with β-glucosidase in high concentrations of various organic solvents (85-99%) was reported as early as 1913 [53].

α-Mannosidase (jack bean) precipitated with ice-cold isopropanol catalyzed the release of *p*-nitrophenyl from Manα-PhNO$_2$-*p* in various solvents (e.g., 99% isopropanol, tetrahydrofuran, acetonitrile) (K. G. I. Nilsson, unpub-

Table 4.13 Influence of Temperature on α-Galactosidase-Catalyzed Formation of Digalactoside

Reaction temperature (°C)	Galα1-3Galα-OMe yield (%)	Galα1-3Galα-OPhNO$_2$-p	
		Yield (%)	Regioselectivity (%)
4	47	22	90
20	43	18	84
40	35	14	78
50	33	13	77

Source: Ref. 91.

lished result). Formation of Manα1-2Manα-OPhNO$_2$-p was detected in up to 97% tetrahydrofuran.

Concentration of Substrates.   An obvious alternative method to decrease the water concentration relative to the acceptor is to increase the concentration of the substrates. Most studies up till now have used far from saturating substrate concentrations, and the molar ratio of acceptor to water has usually been less than 0.01. Still, the ratios of transglycosylation to hydrolysis have been in the range 0.1–1.

The effect of high substrate concentration was shown for the synthesis of Galα1-3Galα-OPhNO$_2$-p from Galα-OPhNO$_2$-p, catalyzed by α-galactosidase (coffee bean). The reaction was carried out at elevated temperature (50°C) to avoid precipitation of the substrates. The yield of α-linked digalactosides increased from 30 to 65% when the initial concentration of Galα-OPhNO$_2$-p increased from 0.15 to 1 M (Nilsson, K.G.I., unpublished data). In another example, with an acceptor of low solubility (GlcNAcβ-OEt-SiMe3), an efficient donor substrate (Galβ-OEtSiMe), was added in excess to minimize secondary hydrolysis and convert most of the acceptor ($\approx$55%) to Galβ1-3GlcNAc-OEtSiMe [83].

In another example, with an acceptor of low solubility (GlcNAcβ-OEt-SiMe3), an efficient donor substrate Galβ-OPhNO$_2$-p was added in excess to minimize secondary hydrolysis and convert most of the acceptor ($\approx$55%) to Galβ1-3GlcNAcβ-OEtSiMe3 [83].

Temperature and pH.   Increased temperature increases the reaction rate but may decrease the yield. In one study, it was found that the yield of disaccharide decreased from 47% at 4°C to 33% at 50°C (Table 4.13) [91]. The yield of transglycosylation reactions also is changed by varying the pH [111].

Immobilization. The effect of modification or immobilization of glycosidases on the yield or regioselectivity has not been extensively investigated. Various glycosidases (α-galactosidase, β-galactosidase, α-mannosidase) were immobilized to tresyl chloride activated agarose and used for synthesis [88]. The yields and regioselectivities were largely unaffected. Accumulation of substrates and/or products inside the carrier may modify the product pattern and the yield [112]. *Aspergillus oryzae* β-galactosidase, modified with polyethylene glycol chlorotriazine derivative, had a higher V_{max} for the amphipatic substrates, GM_3-ganglioside, than the native enzyme [113].

The use of an adsorbent that selectively removes synthesized oligosaccharides may help to reduce secondary hydrolysis in transglycosylation reactions in order to increase product yield. Such systems (with, e.g., active carbon as adsorbent) have been used to increase the yield in equilibrium-controlled oligosaccharide synthesis [114,115].

Equilibrium-Controlled Synthesis (Reversed Hydrolysis)

A. C. Hill, who reported on the synthesis of maltose by reversed hydrolysis from glucose (40%) in 1898, was one of the first to demonstrate enzymatic synthesis in vitro [116]. Despite this early success, few examples of the preparative synthesis of the oligosaccharide structures found in glycoconjugates using revised hydrolysis have been reported.

The formation of glycoside linkages between glycosyl donors (DOH) and various hydroxyl group containing acceptors (HOA) is not favored under physiological conditions.

$$DOH + HOA \rightleftharpoons DOA + H_2O$$

The equilibrium constants for the formation of a few different disaccharides have been reported [99,117,118]. The equilibrium constants (including the water concentration) for the formation of maltose and isomaltose from glucose with glucoamylase were estimated to be 0.28 and 3.10, respectively [117]. Similar differences between the equilibrium constants for the formation of linkage isomers were reported for α-linked dimannose formation catalyzed by jack bean α-mannosidase [0.66, 0.99, and 3.4 for the (α1-2)-, (α1-3)-, and (α1-6)-linked dimannose, respectively] [118]. Recently, the thermodynamics of the enzymatic hydrolysis of cellobiose, gentiobiose, maltose, and isomaltose were studied using HPLC and microcalorimetry [119]. The entropy changes obtained for hydrolysis of these compounds were in the range 32-43 J/mol/K, while the enthalpy change was either positive or negative depending on the disaccharide.

A high substrate concentration has to be used to obtain a reasonable yield of product in equilibrium-controlled synthesis of oligosaccharides, and this is facilitated by using a high temperature. Thus, with 83% (w/w) of mannose, jack bean α-mannosidase catalyzed the formation of about 1 M dimannose at 55°C [118]. Unfortunately, a very complex mixture of different disaccharides and higher oligosaccharides was obtained and none of the products was purified to homogeneity. Oligoglucosides have been prepared analogously [116,120].

Sugars or polyols are known to stabilize enzymes against thermal inactivation [121] and higher temperatures than those allowed at lower substrate concentration could be used in the experiments above. However, many disaccharides have a negative enthalpy of formation (i.e., exothermic reaction) and the yield of the equilibrium-controlled synthesis of these compounds will decrease with increasing temperature. Linkage isomers often have different enthalpies of formation, and a change of temperature may thus change the ratio of formed isomers.

The stereospecificity and the glycon and acceptor specificity of glycosidases are also important. For example, the use of α-glycosidases results in the exclusive formation of α-linked oligosaccharides. Moreover, in a mixture of different monosaccharides, only one will serve as glycosyl donor, provided a pure glycosidase preparation is used. A glycosidase with low acceptor specificity will produce several linkage isomers in amounts determined by the equilibrium concentration.

Isomers not formed to an appreciable extent in transglycosylation reactions, where rate constants may be unfavorable, may be formed in equilibrium-controlled reactions. This has been shown for the formation of maltose and isomaltose with glucoamylase [117] and for the formation of amino acid sugars catalyzed by glycosidases. Thus, while no synthesis was observed with nitrophenyl glycosides as donors and amino acids (serine, threonine, tyrosine) or derivatives thereof as acceptors, synthesis by reversed hydrolysis was observed when high concentrations of either N-acetylgalactosamine, galactose, or mannose were used as donors [122].

The use of an adsorbent that removes the synthesized oligosaccharides from the reaction mixture is a promising approach to increase the yield of reversed hydrolysis. Wallenfels described the use of a carbon–celite column to increase the yield of oligosaccharides produced by β-galactosidase [114]. This type of column can be reused after elution of products with aqueous ethanol. Similarly, the yield of β-galactosidase catalyzed synthesis of (β 1-6)-linker Gal-GlcNAc from galactose and N-acetylglucosamine was 9% when the reaction mixture was circulated through an active carbon column [115]. Reuse of substrates three times increased the yield to 32%.

An obvious advantage of the equilibrium approach is that cheap, unmodified substrates are used. In transglycosylation reactions, the donor is consumed and thus cannot be reused. However, glycosides are better substrates for glycosidases than monosaccharides with a free reducing end and, therefore, higher amounts of enzyme are required in the equilibrium-controlled reactions. In addition, pure enzyme preparations and/or longer reaction times have to be used to avoid the simultaneous formation of, for example, α- and β-glycosides.

Because of the simplicity of the method, reports on the successful equilibrium-controlled synthesis of oligosaccharide structures of glycoconjugates are to be expected. The regioselectivity may be changed by using acceptor glycosides and the use of, for example, methyl glycosides as acceptors, and a second glycosidase for hydrolysis of undesired linkage isomers may reduce the current problems of product isolation. We are investigating the selective synthesis of a complex tetrasaccharide on a larger scale, using one glycosidase for equilibrium-controlled synthesis of a trisaccharide, which is then used as acceptor in a transglycosylation reaction with a second glycosidase.

SYNTHESIS WITH GLYCOSYLTRANSFERASES

Introduction

In a broader sense, glycosyltransferases can be defined as enzymes transferring a glycosyl residue from a donor to an acceptor substrate; such reactions are catalyzed by glycosidases and phosphorylases, in addition to the enzymes classified as glycosyltransferases (EC 2.4) [55]. Indeed, in the 1940s transglycosylation from sugar phosphates was advanced as a possible general mechanism for the biosynthesis of polysaccharides in vivo [123]. However, the role of phosphorylases and glycosidases is now considered to be primarily degradative, while the highly specific glycosyltransferases are considered to be the primary basis for the specific synthesis of the sugar chains produced by a cell [1,124–126].

The glycosyltransferases catalyze the regio- and stereospecific formation of glycosidic linkages between an activated glycosyl donor (a nucleotide sugar, NDP–donor in the reaction scheme that follows) and an acceptor (acceptor = mono- or oligosaccharide or a derivative thereof):

$$\text{NDP-donor + acceptor} \longrightarrow \text{donor-acceptor + NDP}$$

where N is either uridine or guanosine, depending on the monosaccharide transferred. The sialyltransferases use CMP-Neu5Ac.

The glycosyltransferases are classified according to the type of donor substrate and the specificity (Table 4.14) {55]. It has been estimated that at least

Table 4.14   Examples of Glycosyltransferases [55,124]

Glycosyltransferase	Donor	Product formed[a]
N-Acetylgalactosaminyltransferase		
(Fucα1-2)Gal α1-3-N-Acetylgalactosaminyl-transferase	UDP-GalNAc	**GalNAcα1-3**(Fucα1-2)Gal-R
Fucosyltransferase		
β-Galactoside α1-2-Fucosyltransferase	GDP-Fuc	**Fucα1-2**Galβ1-3(4)GlcNAc-R
Galactosyltransferase		
β-N-Acetylglucosaminide β1-4-Galactosyltransferase	UDP-Gal	**Galβ1-4**GlcNAc-R
(Fucα1-2)Gal α1-3-Galactosyltransferase	UDP-Gal	**Galα1-3**(Fucα1-2)Gal-R
Sialyltransferase		
β-Galactoside α2-3-Sialyltransferase	CMP-Neu5Ac	**Neu5Acα2-3**Galβ1-3GalNAc-R
α-N-Acetylgalactosaminide α2-6-Sialyltransferase	CMP-Neu5Ac	Neu5Acα2-6GalNAc1α-Thr/Ser Galβ1-3

[a]Boldface indicates the linkage formed.

100 different glycosyltransferases are required for the synthesis of the complex carbohydrate chains of glycoconjugates [1,124–126].

Of interest for in vitro synthesis of glycoconjugate glycans are the following transferases: fucosyl-, galactosyl-, mannosyl-, N-acetylgalactosaminyl-, N-acetylglucosaminyl-, and sialyltransferases. In these families there are several different enzymes with different linkage and acceptor specificities (Table 4.14) [55].

The high specificity of the reactions and the high yields (up to 95% reported yield) that can be obtained make these catalysts attractive for the synthesis of complex carbohydrates. Preparative synthesis of glycoconjugate glycans with glycosyltransferases has been hampered by their cofactor dependence and low availability. The often high specificity for the structure and sequence of the acceptor reduces the range of acceptors that can be used in synthetic applications, and some of the glycosyltransferases characterized have been reported to be relatively unstable [124]. Availability, specificity, stability, and cofactor dependence in relation to preparative synthesis are discussed below.

Availability

The glycosyltransferases involved in the biosynthesis of glycoconjugates are membrane-bound proteins found in the endoplasmatic reticulum or in the Golgi apparatus and are present in low concentrations (often < 0.1%) [124,127]. Only a minority of the glycosyltransferases have been isolated and properly characterized. Some of the isolated glycosyltransferases are commercially available, but they are expensive and can be obtained only in limited quantities.

Isolation

Glycosyltransferases have been purified with straightforward methods (homogenization, extraction, ion-exchange, and/or size exclusion chromatography and affinity chromatography) to apparent homogeneity and in quantities that allow synthesis on a gram-scale and above (Table 4.15) [127–134].

In addition, soluble glycosyltransferases present in extracellular fluids, including serum and milk, have been identified. Purification of these enzymes does not require detergents and fractionation by simple methods (ammonium sulfate precipitation, anion-exchange and/or affinity chromatography) have afforded preparations with a purity sufficient for synthetic applications [131,132].

The introduction of affinity chromatography has been important in the isolation of homogeneous glycosyltransferases. General ligands have been used as affinity ligands, such as nucleotides (GDP, CDP, UDP) coupled via a spacer (hexylamine) to the chromatographic material [127,130]. Related transferases often have different affinities for the nucleotide ligands, and separation has been achieved by elution with, for example, a salt gradient.

Affinity chromatography of increased efficiency has been achieved by using ternary complex formation between enzyme, nucleotide, and acceptor saccharide bound via a spacer to the support [129,130]. For example, Rosevear et al. achieved a 100,000-fold purification of β-galactoside α1-2-fucosyltransferase with Galβ1-4GlcNAc-hexylamine agarose [133]. Strong binding of the enzyme to the support was obtained when the buffer contained 0.5 mM GMP, and elution was achieved by removal of GMP from the eluting buffer. Recently, the rapid purification of N-acetyl-β-D-glucosaminide α1-3-fucosyltransferase in 35% yield using the same type of affinity ligand was reported [134]. Another example is the (Fucα1-2)Gal α2-3GalNAc-transferase from human serum [128] or human milk [135], which specifically binds to agarose, and this was used for the efficient isolation of the enzyme.

Because of the high donor and acceptor specificity of the glycosyltransferases, it is not always necessary to use a pure enzyme preparation for synthesis. For example, β-galactoside α2-3-sialyltransferase and N-acetyl-

Table 4.15 Availability of Glycosyltransferases

Enzyme	Source	U/kg[a]	U/mg[b]	Ref.
(Fucα1–2)Gal α1–3–N–Acetylgalactosaminyltransferase	Human milk	15	0.5	128
Gal α1–3–Sialyltransferase	Porcine submaxillary gland	10	11	127
Galβ1–4GlcNAc α2–6–Sialyltransferase	Rat liver	53	8	129
α–GalNAc α2–6–Sialyltransferase	Porcine submaxillary gland	25	41	127
Gal α1–2–Fucosyltransferase	Porcine submaxillary gland	370	129	127
GlcNAc β1–4–Galactosyltransferase	Human milk	83	15	130

[a]Activity in tissue.
[b]Specific activity of isolated enzyme.

galactosaminide α2-6-sialyltransferase from porcine submaxillary glands are difficult to separate on CDP–hexanolamine agarose, but the enzymes have different acceptor specificities [124]. The former enzyme is active with Galβ1-3GalNAc as acceptor, but the α2-6-sialyltransferase does not use this acceptor. Thus, for the preparation of Neu5Acα2-3Galβ1-3GalNAc from CMP–Neu5Ac and Galβ1-3GalNAc, separation of the two enzymes is not necessary and a simplified purification scheme can be used [83].

Similarly, human milk α1-4-fucosyltransferase was partially purified by ammonium sulfate fractionation followed by CM–Sephadex and used for the specific synthesis of Neu5Acα2-3Galβ1-3(Fucα1-4)GlcNAc (sialyl-Lewis-a) from GDP–Fuc and Neu5Acα2-3Galβ1-3GlcNAc [131].

However, the general preparation of complex oligosaccharides on a larger scale is not practical with isolated glycosyltransferases. Therefore, the success of genetic engineering in increasing the availability of these enzymes is important. The cDNAs for several glycosyltransferases have been obtained [126,136-138]. It has been shown that by replacing the cDNA sequence coding for the NH2-terminal amino acid sequence, thought to be responsible for membrane anchoring of the glycosyltransferase, with the sequence for a cleavable peptide, secretion of catalytically active enzyme occurred [139,140]. The number of cloned glycosyltransferases is rapidly increasing, and the ready availability of recombinant glycosyltransferases may be a reality in the near future.

Cofactor Dependence

The nucleotide sugars, GDP–Fuc, GDP–Gal, GDP–Man, UDP–GlcNAc, UDP–GalNAc, and CMP–Neu5Ac, which are used as glycosyl donors in the synthesis of complex carbohydrates, are commercially available but are prohibitively expensive for large-scale synthesis. Simple enzymatic methods for the preparative-scale synthesis of UDP–Gal [141], CMP–Neu5Ac [142-148], and recently UDP–GlcNAc [149] have been developed. Enzymatic synthesis in vitro of other important nucleotide sugars has been described [150].

An immobilized multienzyme system for the production of Galβ1-4GlcNAc with GlcNAc β1-4-galactosyltransferase produced the required nucleotide sugar UDP–Gal in situ and recycled the cofactor UDP [141].

CMP–Neu5Ac has been synthesized from CTP and Neu5Ac with soluble or immobilized CMP–Neu5Ac synthase (EC 2.7.7.43) from various sources (e.g., frog liver, bovine submaxillary glands, calf brains, bovine liver) (Fig. 4.8). The enzyme has been cloned. To obviate the need for the expensive substrates CTP and N-acetylneuraminic acid, multienzyme systems starting from cheaper substrates have been devised [147,148]. In one system, CMP–Neu5Ac was prepared from N-acetylglucosamine, pyruvate, CMP, and phosphoenol-

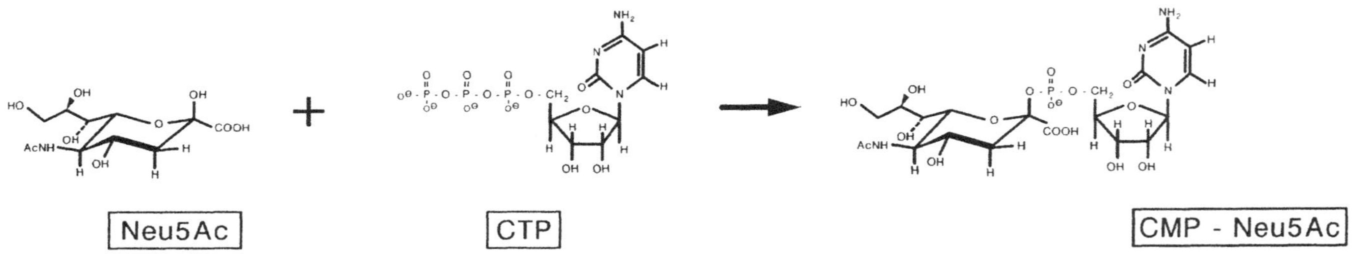

Figure 4.8 Synthesis of CMP–Neu5Ac from CTP and *N*-acetylneuraminic acid using CMP–sialate synthase.

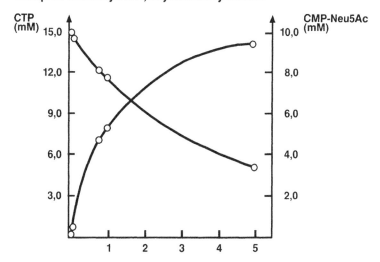

Figure 4.9   Synthesis of CMP-Neu5Ac as a function of time with an immobilized porcine liver CMP-sialate synthase preparation. (From ref. 151.)

pyruvate [148]. However, higher yields of CMP–Neu5Ac were obtained by running the reactions separately. A simple purification procedure for the isolation of CMP–Neu5Ac based on precipitation with ethanol–NH4OH was described [148].

In another study, a stable and highly active preparation of CMP–Neu5Ac synthase was obtained in a few hours from bovine liver by immobilizing a crude extract to tresyl-activated agarose [151]. This preparation was used for repeated synthesis of CMP–Neu5Ac from CTP and Neu5Ac (yield > 80%) with low or no loss of activity. The isolation of CMP–Neu5Ac was facilitated by the immobilization, since the reaction mixture was not contaminated with the various substances otherwise present in crude extracts. The reaction was followed by FPLC, which allowed the reaction to be followed with time (Fig. 4.9).

The CMP–Neu5Ac synthase from rat is tolerant to variations of *N*-acetylneuraminic acid at C–4, C–7, and C–9 and at the *N*-acyl group, whereas 8-OH, 2-OH, and the NH of the NHAc group seem to be responsible for recognition of the sialic acid derivative by the enzyme [152]. A number of CMP–Neu5Ac derivatives have been enzymatically prepared from CTP and analogues of *N*-acetylneuraminic acid [143,153,154]. In one study, C–9 analogues were readily incorporated into asialo-α-acid glycoprotein by Galβ1-4GlcNAc α2-6-sialyltransferase from rat liver (EC 2.4.99.1) [153]. This technique should be useful for the preparation of sialoglycoconjugates with modified biological properties, such as sialidase resistance (9–amino–Neu5Acα-glycosides are not hydrolyzed by sialidases) [153].

Specificity

The synthesis of oligosaccharides with glycosyltransferases is advantageous compared with glycosidases because only one regioisomer is formed and quantitative yields are easily obtained. For synthetic purposes, it is of interest to know the range of structures that can be used in synthesis. Glycosyltransferases are considered to be specific for the nucleotide sugar and the acceptor sugar.

A number of exceptions are known, such as the (Fucα1-2)Gal α1-3NAc-transferase, which has some capacity to transfer galactose in addition to N-acetylgalactosamine, and the enzyme thus has the ability to synthesize both the A and B blood group determinants [155]. Moreover, the GlcNAc α1-3Fuc-transferase from human milk transfers fucose to both glucose and N-acetylglucosamine [156]. Recently, the Galβ1-4GlcNAc α2-6-sialyl-transferase from rat liver was shown to use oligosaccharides of type Manβ1-4GlcNAc-R as acceptors to form Neu5Acα2-6Manβ1-4GlcNAc-R [157]. This contrasts with the reported high specificity of the enzyme for galactose when (β1-4)-linked to GlcNAc-R. Deoxy sugars have also been used as acceptors [158].

The list of such exceptions may be much longer when more transferases are available for thorough characterization. As mentioned above, the α2-6-sialyltransferase from rat liver [159] was found to transfer analogues of Neu5Ac [153], and other reports have described the transfer of analogues of UDP-Glc [160], UDP-Gal [161], and GDP-Man [162] to acceptor sugars. It was recently shown that CMP-9-fluoresceinyl-Neu5Ac was a suitable donor substrate for five different sialyltransferases [163].

Examples of glycosyltransferases with high acceptor sequence specificity are the α1-3-Gal and α1-3-GalNAc-transferases, which both require the nonreducing terminal sequence Fucα1-2Gal [124,164,165]. The branch-generating N-acetylglucosaminyltransferases (GnT) are able to recognize specifically the correct branch position of their acceptor; GnT- I selects one of five possible mannose units as acceptor [166].

Other glycosyltransferases often show lower acceptor specificity but have preferences for branch localization, degree of branching, and the primary sequence of the acceptor (Table 4.16). The porcine β-galactoside α2-3-sialyltransferase shows a preference for acceptors with the sequence Galβ1-3GalNAc-R, but Galβ1-3GlcNAc-R is also active [167], and the enzyme has been used for sialylation of these structures [83]. The

Table 4.16 Acceptor Specificity of Three Different Glycosyltransferases

Acceptor	Gal α2-3-Sialyltransferase[a]		Galβ1-4GlcNAc α2-6-Sialyltransferase[b] (relative rate)	Gal α1-2-Fucosyltransferase[c]	
	K_m	v_{max}		K_m	V_{max}
Galβ1-3GlcNAc	65	4.4	1	0.63	98
Galβ1-4GlcNAc	42	0.11	100	2.11	30
Galβ1-6GlcNAc	29	0.24	0	2.56	21
Galβ1-3GalNAc	0.21	8.9	0	0.28	86
Galβ1-6GalNAc	No/low activity		1	1.95	18
Galβ1-4Glc	180	0.62	1	10.7	81
Galβ-OMe	No/low activity		n.d.	18	96

[a]Procine submaxillary gland, data from Ref. 124.
[b]Rat liver, data from Ref. 159.
[c]Procine submaxillary gland, data from Ref. 127.

α1-3-fucosyltransferase from human liver transfers fucose to the O-3 position of the GlcNAc moiety of Galβ1-4GlcNAc and Neu5Acα2-3Galβ1-4GlcNAc, but Galβ1-3GlcNAc and Neu5Acα2-6Galβ1-4GlcNAc are not acceptors (or poor ones) [168].

The best studied glycosyltransferase and the first to be purified to homogeneity, GlcNAc β1-4Gal-transferase [169], has a broad tolerance to different acceptors and, in addition to glucose or N-acetylglucosamine, a number of oligosaccharide sequences containing GlcNAc are good acceptors as well as GlcNAc and GlcNAc-containing oligosaccharides coupled to solid phases [132,170,171] (Table 4.17). In the mammary gland the acceptor specificity is modulated by a protein, α-lactalbumin, which inhibits the transfer of UDP-Gal to GlcNAc and stimulates the synthesis of lactose [172,173]. It is possible that other proteins may be found that modulate the specificity of glycosyltransferases, thus broadening the range of acceptors that can be used in synthesis.

A pure enzyme preparation is, of course, required to establish the specificity. The Lewis-a gene associated α-fucosyltransferase from human milk was considered to be a possible α3/4-transferase, which transfers L-fucose to both the O-3 and O-4 positions of GlcNAc oligosaccharides. However, more extensive purification indicated the presence of two different fucosyltransferases [174].

Examples of Preparative Synthesis

A number of oligosaccharide sequences common in glycoconjugates have been prepared on the milligram-gram scale, employing various glycosyltransferases for the final glycosylation steps (Table 4.17) [83,131-133,141,170,171,175-185].

Several of the products were prepared in amounts sufficient for characterization using the routine methods used in organic chemistry (NMR spectroscopy, optical rotation, elemental analysis and, if crystalline, melting points). The high specificity of glycosyltransferases facilitated product isolation by column chromatography (usually ion-exchanger or silica and molecular sieve for desalting). Some of the structures were synthesized in vitro for the first time.

Some of the reactions can be scaled up considerably. For example, the trisaccharides Galβ1-4GlcNAcβ1-3Gal and Galβ1-4GlcNAcβ1-6Gal were synthesized on a millimolar scale in 2 days (70% yield) with about 2 units of GlcNAc β1-4-galactosyltransferase isolated from 20 mL of human milk [175]. An immobilized multienzyme system similar to that previously described for the synthesis of Galβ1-4GlcNAc [141] was used, and the synthesis could be repeated with negligible loss of enzyme activity. Thus, starting from a few liters of human milk, synthesis on a molar scale should be possible. Other transferases are less abundant; in the case of sialyltransferases the scale of synthesis was 10–50 mg, which however, easily can be scaled up 10 to 1000-fold.

The glycosyl donor is often used in excess over the acceptor, if abundant acceptors like the simple mono- or oligosaccharides (e.g., lactose) or acceptors of low reactivity are not used. In most of the reported reactions for preparative purposes, a relatively low substrate concentration (5–50 mM) has been used. The use of higher concentrations is hindered by substrate and product inhibition. Thus, the bovine N-acetylglucosaminide β1–4-galacto-syltransferase is inhibited by UDP-Gal, and the concentration of the donor in reactions with this transferase is usually kept at 10 mM or lower [132,133]. Product inhibition may be reduced by the use of additional enzymes, such as alkaline phosphatase for removal of, for example, CMP in sialyltransferase-catalyzed reactions. Some glycosyltransferases, such as α1–2-fuco-syltransferase [133], have been proposed to contain an intrinsic hydrolase activity, which lowers the yield of product.

The obvious advantage compared with chemical synthesis is that specific synthesis is achieved in one reaction step, with good yield (reported yields of 30–95%) and under mild conditions (pH 6–8, 20–37°C). The latter property is especially important with the relatively labile N-acetylneuraminic acid glycosidic linkages.

Several sialylated structures that appear as terminal sequences of glycoproteins and glycolipid oligosaccharides, as well as common sequences found at branching points (GlcNAcβ1–6) and as repeating units in complex oligosaccharides, have been synthesized. Some of the synthesized structures are cancer-associated antigens, such as Neu5Acα2–3Galβ1–3GlcNAc and Neu5Acα2–3Galβ1–3(Fucα1–4)GlcNAc (Sia-Le-a) [5,10], or are receptors for toxins, viruses, or bacteria.

Various acceptor glycosides (–Me, –Ph, –(CH2)5NH2, –(CH2)8COOMe, –EtBr) have been used to produce products suitable for various applications, such as inhibition studies, as enzyme substrates, or for attachment to proteins or chromatographic material. Synthesis of glycopeptides is also possible, as shown in Table 4.17.

Most of the tri- and higher oligosaccharide structures were prepared by means of a combination of chemical and enzymatic syntheses, using chemical methods for the synthesis of the smaller acceptor structures and glycosyltransferases for the final glycosylation step. Because of the complexity of chemical synthesis, even for preparation of di- or trisaccharides (usually 5–7 and 8–15 reaction steps, respectively), it is desirable to substitute enzymatic methods for the remaining chemical steps. Rosevear et al. used GlcNAc β1–4-galactosyltransferase and porcine α1–2-fucosyltransferase for the se-

Table 4.17 Examples of Synthesis of Glycoconjugate Oligosaccharide Sequences with Glycosyltransferases

Enzyme	Product[a]	Ref.
N–Acetylglucosaminyltransferase V	**GlcNAcβ1–6**(GlcNAcβ1–2)Manα1–6Manβ–O(CH$_2$)$_8$COOMe	182
Gal α1–3/4–Fucosyltransferase	**Fucα1–2**Galβ–OEt	133
	Fucα1–2 Galβ1–4GlcNAc	133
	Fucα1–2 Galβ–O(CH$_2$)$_6$NH$_2$	133
GlcNAc α1–3/4–Fucosyltransferase	Neu5Acα2–3Galβ1–3GlcNAcβ1–3Galβ–O(CH$_2$)$_8$COOMe	131
	| **Fucα1–4**	
	Galβ1–3GlcNAcβ–O(CH$_2$)$_8$COOMe	183
	| **Fucα1–4**	
GlcNAc β1–4–Galactosyltransferase	**Galβ1–4**GlcNAc	132,141
	Galβ1–4 GlcNAcβ–O(CH$_2$)$_6$NH$_2$	132
	Galβ1–4 Glcβ–O(CH$_2$)$_6$NH–agarose	132
	Galβ1–4 Glcβ–OCH$_2$C$_6$H$_4$NO$_2$–CONH–polymer	170
	Galβ1–4 GlcNAcβ1–3Gal	175,178
	Galβ1–4 GlcNAcβ1–6Gal	175,178
	GlcNAcβ1–3Galβ1–4Glcβ–OMe	176,178
	| **Galβ1–4**GlcNAcβ1–6	

Fucα1-6GlcNAcβ O(CH₂)₈COOMe — 180

      |

   Galβ1-4

Galβ1-4 GlcNAcβ1-L-Asn — 184,185

Galβ1-4 GlcNAcβ1-4GlcNAcβ1-L-Asn — 185

Gal α2-3-Sialyltransferase

Neu5Acα2-3 Galβ1-3GalNAcα-OEt — 83

Neu5Acα2-3 Galβ1-3GalNAcβ-OEtBr — 83

Neu5Acα2-3 Galβ1-3GalNAcβ-O(CH₂)₅COOMe — 177

Neu5Acα2-3 Galβ1-3GlcNAcβ-OMe — 83

Neu5Acα2-3 Galβ1-3GlcNAcβ-O(CH₂)₈COOMe — 131

Gal β1-3/4GlcNAc α2-3-Sialyltransferase

Neu5Acα2-3 Galβ1-3GlcNAcβ-OMe — 177

Neu5Acα2-3 Galβ1-4GlcNAcβ-OMe — 177

Neu5Acα2-3 Galβ1-4GlcNAcβ-O(CH₂)₅COOMe — 177

Neu5Acα2-3 Galβ1-4Glcβ-OMe — 177

Neu5Acα2-3 Galβ1-4GlcNAcβ1-3Galβ1-4Glc — 177

Gal β1-4GlcNAc α2-6-Sialyltransferase

Neu5Acα2-6 Galβ-OMe — 177

Neu5Acα2-6 Galβ1-4GlcNAc — 148,177,181

Neu5Acα2-6 Galβ1-4GlcNAcβ1-L-Asn — 184

Neu5Acα2-6 Manβ1-4GlcNAcβ1-4GlcNAc — 157

GalNAc α2-6-Sialyltransferase

Neu5Acα2-6 — 179

    |

Neu5Acα2-3Galβ1-3GalNAcβ-OPh

ªBoldface indicates linkage formed.

quential synthesis of Galβ1-4GlcNAc (95% yield) and Fucα1-2Galβ1-4GlcNAc (83% yield) [133]. Thiem and Treder [181] used the abundant galactosyltransferase and the cofactor regeneration system described by Wong et al. [141] for synthesis of Galβ1-4GlcNAc, which was then sialylated with α2-6-sialyltransferase (52% yield).

The general use of methods entirely based on glycosyltransferases is limited by the availability of glycosyltransferases and nucleotide sugars.

In an alternative approach, the more readily available glycosides are used to produce acceptor oligosaccharides, and the use of glycosyltransferases is limited to the final glycosylation steps when the demand on regiospecificity is higher [83]. β-Galactosidase from bovine testes and β-galactoside α2-3-sialyltransferase from porcine submaxillary glands were used for the synthesis of the methyl glycoside of the cancer-associated sequence Neu5Acα2-3Galβ1-3GlcNAcβ- and the ethyl glycoside of the common terminal sequence Neu5Acα2-3Galβ1-3GalNAcα- in glycolipids and O-linked glycoproteins (Fig. 4.10) [83]. The 2-bromoethyl glycoside of the latter structure was also synthesized using the alternative acceptor GalNAcβ-OEtBr (prepared with β-hexosaminidase from 2-bromoethanol and GalNAcβ-OPNO$_2$-p, the yield was 55%).

Although not optimized, the yields of α2-3-sialylated Galβ1-3GalNAcα-OEt and Galβ1-3GlcNAcβ-OMe were 72 and 30%, respectively, a difference that can be attributed to the higher V_{max} and lower K_m shown by the enzyme for the former acceptor (Table 4.16).

Instead of the chemoenzymatic approach used for synthesis of several of the structures in Table 4.17, one can substitute the combined use of glycosidase and glycosyltransferase (cf. Fig. 4.10).

Stability

Several of the glycosyltransferases that have been used for preparative synthesis show good stability. The GlcNAcβ1-4-galactosyltransferase from bovine milk can be stored for at least 2 years at 4°C, and partially purified porcine β-galactoside α1-2-fucosyltransferase was stable at 4°C for at least 3 months [133]. The β1-4-galactosyltransferase from bovine milk have been immobilized by various methods and used for repeated synthesis and with minimal loss of activity [141,175].

A commercial preparation of porcine β-galactoside α2-3-sialyltransferase was immobilized to tresyl-activated agarose with high retention of activity (80%) and under mild conditions (pH 7.5, room temperature, 2 h) [146]. Similarly, a freeze-dried, partially purified preparation of the same sialyltransferase was coupled to tresyl-activated agarose with 55% retention of added activity and used for repeated synthesis (three times) with negligible or

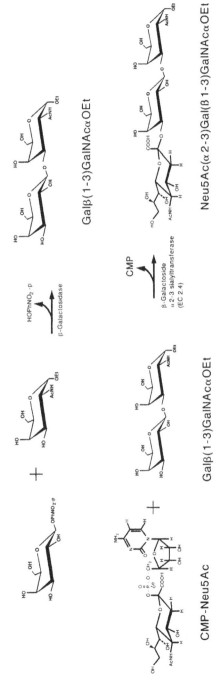

Figure 4.10   Enzymatic synthesis of a trisaccharide by the sequential use of a glycosidase (β-galactosidase) and a glycosyltransferase (β-galactoside $\alpha2$–3–sialyltransferase) [83].

no loss of activity [151]. The use of immobilized enzyme obviated the need for detergent, which was used in earlier syntheses with the soluble enzyme [177].

CONCLUSION

Enzymatic or chemoenzymatic methods for the preparation of a number of glycoconjugate oligosaccharide sequences active as blood group antigens, cancer-associated antigens, or cell surface receptors are now available. Several or most of the disaccharide sequences found in glycoproteins or glycolipids, as well as a few trisaccharide sequences, have been prepared in simple, stereospecific reactions employing abundant glycosidases as catalysts. Structures have been prepared on the gram scale and above with these methods.

Higher oligosaccharides can be prepared by a combination of chemical and enzymatic methods or by the sequential use of glycosyltransferases or of glycosidase and glycosyltransferase. Oligosaccharide analogues and glycosides with aglycons suitable for various purposes can be prepared.

Future development will include the combined use of exoglycosidases and endoglycosidases for the synthesis of higher structures. In addition to the screening of suitable glycosidases, genetic engineering of glycosidases should be useful to improve yield and regioselectivity.

The development of recombinant techniques and efficient isolation procedures to increase the availability of glycosyltransferases is required for the wider application of these enzymes in the preparative synthesis of glycoconjugate glycans. The number of cloned glycosyltransferases is rapidly increasing.

Glycosyltransferases, O- and N-glycosidases, and other enzymes that are active on carbohydrates or carbohydrate derivatives (e.g., lipases) are likely to be used in elaborate enzyme reaction systems, alone or in combination with chemical methods for the modification of glycoproteins and glycolipids and for the synthesis of glycoconjugate carbohydrate sequences, analogues thereof, glycopeptides, and other glycosides suitable for clinical and other applications.

REFERENCES

1. Ginsburg, V., and Robbins, P. W. (eds.). *Biology of Carbohydrates*, Vol 2, Wiley, New York, 1984.
2. Brady, R. (ed.). *Chemistry and Physics of Lipids*, Vol. 42, Elsevier, New York, 1986.
3. Rademacher, T. W., Parekh, R. B., and Dwek, R. A. Glycobiology, *Ann. Rev. Biochem. 57*, 785–838 (1988).

4. Sharon, N., and Lis, H. Lectins as cell recognition molecules, *Science, 246,* 227-234 (1989).
5. Hakomori, S. Tumor-associated carbohydrate antigens, *Annu. Rev. Immunol. 2,* 103-126 (1984).
6. Karlsson, K.-A. Animal glycosphingolipids as membrane attachment sites for bacteria, *Annu. Rev. Biochem. 58,* 309-350 (1989).
7. Hart, G. W., Holt, G. D., and Haltiwanger, R. S. Nuclear and cytoplasmic glycosylation: Novel saccharide linkages in unexpected places, *Trends Biochem. Sci. 13,* 380-384 (1988).
8. Ashwell, G., and Hartford, J. *Annu. Rev. Biochem. 51,* 531-554 (1982).
9. Paulson, J.C. Glycoproteins: What are the sugar chains for? *Trends Biochem. Sci. 14,* 272-276 (1989).
10. Holmgren, J. (ed.). *Tumor Marker Antigens,* Studentlitteratur, Lund, Sweden, 1985.
11. Livingston, P. O., Natoli, E. J., Calves, M. J., Stockert, E., Oettgen, H. F., and Old, L. J. Vaccines containing purified GM2 ganglioside elicit GM2 antibodies in melanoma patients, *Proc. Natl. Acad. Sci. USA, 84,* 2911-2915 (1987).
12. Kobata, A. Structures of sialic acid-containing *N*-glycans in cancer cells. In *Sialic Acids 1988* (R. Schauer and T. Yamakawa, eds.), Kieler Verlag, Kiel, 1988, pp. 206-207.
13. Paulsen, H. Synthesis of complex oligosaccharide chains of glycoproteins, *Chem. Soc. Rev. 13,* 15-45 (1984).
14. Schmidt, R. R. Neue Methoden zur Glykosid und Oligosaccharidsynthese—Gibt es Alternativen zur Koenigs-Knorr-Methode? *Angew. Chem. 98,* 213-236 (1986).
15. Lemieux, R. U. The origin of the specificity in the recognition of oligosaccharides by proteins, *Chem. Soc. Rev. 18,* 347-374 (1989).
16. Karlsson, K.-A. Animal glycolipids as attachment sites for microbes, *Chem. Phys. Lipids, 42,* 153-172, (1986).
17. Lemieux, R., Bock, K., Delbaere, L. T. J., Koto, S., and Rao, V. S. The conformations of oligosaccharides related to the ABH and Lewis human blood group determinants, *Can. J. Chem. 58,* 631-652 (1980).
18. Hakomori, S. Glycosphingolipids in cellular interaction, differentiation and oncogenesis, *Annu. Rev. Biochem. 50,* 733-764 (1981).
19. Kagawa, Y., Takasaki, S., Utsumi, J., Hosoi, K., Shimizu, H., Kochibe, N., and Kobata, A. Comparative study of the asparagine-linked sugar chains of natural human interferon-β1 and recombinant human interferon-β1 produced by three different mammalian cell lines, *J. Biol. Chem. 263,* 17508-17515 (1988).
20. Tsuda, E., Goto, M., Murkami, A., Akai, K., Ueda, M., Kawanish, G., Takahashi, N., Sasaki, R., Chiba, H., Ishihara, H., Mori, M., Tejima, S., Endo, S., and Arata, Y. Comparative structural study of *N*-linked oligosaccharides of urinary tract and recombinant erythropoietins, *Biochemistry, 27,* 5646-5654 (1988).
21. Carr, S. A., Hemling, M. E., Folena-Wasserman, G., Sweet, R. W., Anumula, K., Barr, J. R., Huddleston, M. J., and Taylor, P. Protein and carbohydrate structural

analysis of a recombinant soluble CD4 receptor by mass spectroscopy, *J. Biol. Chem. 264*, 21286-21295 (1989).

22. Conradt, H. S., and Nimtz, M. Recombinant human secretory glycoproteins: Cell type and glycosylation site dependence of oligosaccharide processing, *Proceedings of the Xth International Symposium on Glycoconjugates*, Jerusalem, 1989, pp. 237-238.

23. Berger, E. G., Greber, U., and Mosbach, K. Galactosyltransferase-dependent sialylation of complex and *endo-N*-acetylglucosaminidase H-treated core *N*-glycans in vitro, *FEBS Lett. 203*, 64-68 (1986).

24. Peterson, C. B., and Blackburn, M. N. Isolation and characterisation of an antithrombin III variant with reduced carbohydrate content and enhanced heparin binding, *J. Biol. Chem. 260*, 610-615 (1985).

25. Dang, C. V., Shin, C. K., Bell, W. R., Nagaswami, C., and Weisel, J. W. Fibrinogen sialic acid residues are low affinity calcium-binding sites that influence fibrin assembly, *J. Biol. Chem. 264*, 15104-15108 (1989).

26. Hunt, R. C., Rigler, R., and Davis, A. A. Changes in glycosylation alter the affinity of the human transferrin receptor for its ligand, *J. Biol. Chem. 264*, 9643-9648 (1989).

27. Wasserman, P. M. *Annu. Rev. Cell Biol. 3*, 109-142 (1987).

28. Fenderson, B. A., Zehavi, U., and Hakomori, S. A multivalent lacto-*N*-fucopentaose III-lysyllysine conjugate decompacts preimplantation mouse embryos, while the free oligosaccharide is ineffective, *J. Exp. Med. 160*, 1591-1596 (1985).

29. Siegelman, M. H., Van de Rijn, M., and Weissman, I L. Mouse lymph node homing receptor cDNA clone encodes a glycoprotein revealing tandem interaction domains, *Science, 243*, 1165-1172 (1989).

30. Drickamer, K. Two distinct classes of carbohydrate-recognition domains in animal lectins, *J. Biol. Chem. 263*, 9557-9560 (1988).

31. Hanai, N., Dohi, T., Nores, G. A., and Hakomori, S. A novel ganglioside, de-*N*-acetyl-GM$_3$ (II^{3}NeuNH$_2$LacCer), acting as a strong promoter for epidermal growth factor receptor kinase and as a stimulator for cell growth, *J. Biol. Chem. 263*, 6296-6301 (1988).

32. Storey, E. L., Anderson, G. J., Mack, U., Powell, L. W., and Halliday, J. W. Desialylated transferrin as a serological marker of chronic excessive alcohol ingestion, *Lancet, 1*, 1292-1293 (1987).

33. Feizi, T. Demonstration by monoclonal antibodies that carbohydrate structures of glycoproteins and glycolipids are onco-developmental antigens, *Nature, 314*, 53-58 (1985).

34. Dennis, J. W., Laferté, S., Waghorne, C., Breitman, M. L., and Kerbel, R. S. β 1-6 Branching of Asn-linked oligosaccharides is directly associated with metastasis, *Science, 236*, 582-585 (1987).

35. Race, R. R., and Sanger, R. *Blood Groups in Man*, 6th ed., Blackwell, Oxford, 1975.

36. Nilsson, K. G. I. Enzymatic synthesis of oligosaccharides, *Trends Biotechnol. 6*, 256-264 (1988).

37. Toone, E. J., Simon, E. S., Bednarski, M. D., and Whitesides, G. M. Enzyme-catalyzed synthesis of carbohydrates, *Tetrahedron Rep. 45*, 5365-5422 (1989).

38. Hakomori, S. Tumor-associated glycolipid markers: Possible targets for drug and immunotoxin delivery, in *Targeting of Drugs with Synthetic Systems* (G. Gregoriadis and G. Poste, eds.), Plenum Press, New York, 1986, pp. 25-40.

39. Falent-Kwast, E., Kovac, P., Bax, A., and Glaudemans, C. P. J. *Carbohydr. Res. 145*, 332-340 (1986).

40. Ekborg, G., Vranesić, B., Bhattacharjee, A. K., Kovác, P., and Glaudemans, C. P. Synthesis of *p*-nitrophenyl β-glycosides of (1-6)-β-D-galactopyranosyl oligosaccharides, *Carbohydr. Res. 142*, 203-211 (1985).

41. Dahmén, J., Frejd, T., Magnusson, G., Noori, G., and Carlström, A. S. Synthesis of spacer-arm, lipid, and ethyl glycosides of the terminal trisaccharide portion of the blood group P_1 antigen: Preparation of neoglycoproteins, *Carbohydr. Res. 129*, 63-71 (1984).

42. Nilsson, B., Larsson, A.-L., Pålsson, Ögren, Y., Back, M., and Nilsson, K. An amplification technology for improving sensitivity when measuring components in biological samples, *J. Immunol. Methods, 108*, 237-244 (1988).

43. Aronson, M., Medalia, O., Schori, L., Mirelman, D., Sharon, N., and Ofek, I. Prevention of colonisation of the urinary tract of mice with *E. coli* by blocking of bacterial adherence with methyl α-D-mannopyranoside, *J. Infect. Dis. 139*, 329-332 (1979).

44. Svanborg-Edén, C., Freter, R., Hagberg, L., Hull, R., Leffler, H., and Schoolnik, G. Inhibition of experimental ascending urinary tract infection by an epithelial cell-surface receptor analogue, *Nature, 298*, 560-562 (1982).

45. Weis, W., Brown, J. H., Cusack, S., Paulson, J. C., Skehel, J. J., and Wiley, D. C. Structure of the influenza virus haemagglutinin complexed with its receptor sialic acid, *Nature, 333*, 426-431 (1988).

46. Crawley, S. C., Hindsgaul, O., Alton, G., Pierce, M., and Palcic, M. M. An enzyme-linked immunosorbent assay for *N*-acetylglucosaminyltransferase-V, *Anal. Biochem. 185*, 112-117 (1990).

47. Kovác, P., Glaudemans, C. P. J., and Taylor, R. B. An efficient unambiguous synthesis of methyl 3-*O*-β-D-galactopyranosyl-β-D-galactopyranoside. Further studies on the specificity of antigalactopyranan monoclonal antibodies, *Carbohydr. Res. 142*, 158-164 (1985).

48. Marra, A., and Sinay, P. A novel stereoselective synthesis of *N*-acetyl-α-neuraminosylgalactose disaccharide derivatives, using anomeric *S*-glycosyl xanthates, *Carbohydr. Res. 195*, 303-308 (1990).

49. Ijima, H., and Ogawa, T. Total synthesis of 3-*O*-(2-acetamido-6-*O*-(*N*-acetyl-α-D-neuraminyl)-2-deoxy-α-D-galactosyl)-L-serine and stereoisomer, *Carbohydr. Res. 172*, 183-193 (1988).

50. Paulsen, H., and Dessen, U. Synthese von *O*-(5-Acetamido-3,5-didesoxy-D-glycero-α-D-galacto-2-nonulopyranosylonsäure)-(2-3)-*O*-β-D-galactopyrano-syl-(1-3)-2-acetamido-2-desoxy-D-galactopyranose, *Carbohydr. Res. 175*, 283-293 (1988).

51. Flowers, H. M., and Sharon, N. Glycosidases, properties and application to the

study of complex carbohydrates and cell surfaces, *Adv. Enzymol.* *48*, 29-95 (1979).

52. Fischer, E. Einfluss der Configuration auf die Wirkung der Enzyme, *Ber. Chem. Ges.* *27*, 2985-2993 (1894).

53. Bourquelot, E., and Bridel, M. Synthèse des glucosides d'alcools à l'aide de l'émulsine et réversibilité des action fermentaires, *Ann. Chim.* *29*, 145-218 (1913).

54. Kobata, A. Use of endo- and exoglycosidases for structural studies of glycoconjugates, *Anal. Biochem.* *100*, 1-14 (1979).

55. *Enzyme Nomenclature*, Academic Press, New York, 1984.

56. Wallenfels, K., and Weil, R. β-Galactosidase, *Enzymes,* *7*, 617-663 (1972).

57. Nisizawa, K., and Hashimoto, Y. Glycoside hydrolases and glycosyltransferases, in *The Carbohydrates, Chemistry and Biochemistry*, 2nd ed. (Pigman, W., Horton, D., and Herp, A., eds.), Academic Press, New York, 1975, pp. 241-301.

58. Phillips, D. C. The hen egg-white lysozyme molecule, *Proc. Natl. Acad. Sci. USA,* *57*, 484-495 (1967).

59. Air, G. M., and Laver, W. G. The neuraminidase of influenza virus, *Proteins,* *6*, 341-356 (1989).

60. Sinnot, M. L. Ions, ion-pairs and catalysis by the lacZ β-galactosidase of *E. coli*, *FEBS Lett.* *94*, 1-9 (1978).

61. Blake, C. C. F., Johnson, L. N., Mair, G. A., North, A. C. T., Phillips, D. C., and Sarma, V. R. Crystallographic studies of the activity of hen egg-white lysozyme, *Proc. R. Soc. London, B167*, 378-388 (1967).

62. Vernon, C. A. The mechanism of hydrolysis of glycosides and their relevance to enzyme catalysed reactions, *Proc. R. Soc. London, B167*, 389-401 (1967).

63. Dey, P. M., and Campillo, E. Biochemistry of the multiple forms of glycosidases in plants, *Adv. Enzymol.* *56*, 141-250 (1984).

64. Dey, P. M., and Pridham, J. B. Biochemistry of α-galactosidases, *Adv. Enzymol.* *36*, 91-130 (1972).

65. Li, S.-C., and Li, Y.-T. Studies on the glycosidases of jack bean meal, *J. Biol. Chem.* *245*, 5153-5160 (1970).

66. Elbein, A. D., Adya, S., and Lee, Y. C. Purification and properties of β-mannosidase from *Aspergillus niger, J. Biol. Chem.* *252*, 2026-2031 (1977).

67. Sinnot, M. L. β-Galactosidase-catalyzed hydrolysis of β-D-galactopyranosyl azide, *Biochem. J.* *125*, 717-719 (1971).

68. Yasawa, S., Madiyalakan, R., Chawda, R. P., and Matta, K. L. α-L-Fucosidase from *Aspergillus niger*: Demonstration of a novel αL-(1-6)-fucosidase acting on glycopeptides, *Biochem. Biophys. Res. Commun.* *136*, 563-569 (1986).

69. Dicioccio, R. A., Klock, P. J., Barlow, J. J., and Matta, K. L. Rapid procedures for determination of *endo-N*-acetyl-α-D-galactosaminidase in *Clostridium prefringens*, and of the substrate specificity of *exo-β*-D-galactosidases, *Carbohydr. Res.* *81*, 315-322 (1980).

70. Distler, J. J., and Jourdian, G. W. The purification and properties of β-galactosidase from bovine testes, *J. Biol. Chem.* *19*, 6772-6780 (1973).

71. Tai, T., Yamashita, K., Ogata-Arakawa, M., Koide, N., Muramatsu, T., Ishiwata, S., Inoue, Y., and Kobata, A. Structural studies of two ovalbumin glycopeptides in relation to the *endo-βN*-acetylglucosaminidase specificity, *J. Biol. Chem.* *250*, 8569-8575 (1975).

72. Yamashita, K., Ischishima, E., Arai, M., and Kobata, A. An α-mannosidase from *Aspergillus saitoi* is specific for α1,2 linkages, *Biochem. Biophys. Res. Commun.* *96*, 1335-1342 (1980).

73. Dicioccio, R. A., Barlow, J. J., and Matta, K. L. Substrate specificity and other properties of αL-fucosidase from human serum, *J. Biol. Chem.* *257*, 714-718 (1982).

74. Van den Eijnden, D. H., Blanken, W. M., and van Vliet, A. Branch specificity of β-D-galactosidase from *E. coli, Carbohydr. Res.* *151*, 329-335 (1986).

75. Corfield, A. P., Wember, M., Schauer, R., and Rott, R. The specificity of viral sialidases, *Eur. J. Biochem.* *124*, 521-525 (1982).

76. Dey, P. M., Campillo, E. M., and Pont Lezica, R. Characterisation of a glycoprotein α-galactosidase from lentil seeds (*Lens culinaris*), *J. Biol. Chem.* *258*, 923-929 (1983).

77. Steers, E., Jr., Cuatrecasas, P., and Pollard, H. B. The purification of β-galactosidase from *E. coli* by affinity chromatography, *J. Biol. Chem.* *246*, 196-200 (1971).

78. Cuatrecasas, P. Purification of neuraminidases (sialidases) by affinity chromatography, *Methods Enzymol.* *28*, 897-902 (1972).

79. Barker, R., Chiang, C. K., Trayer, I. P., and Hill, R. L. Monosaccharides attached to agarose, *Methods Enzymol.* *34*, 317-328 (1974).

80. Veide, A., Smeds, A.-L., and Enfors, S.-O. A process for large-scale isolation of β-galactosidase from *E. coli* in an aqueous two-phase system, *Biotechnol. Bioeng.* *25*, 1789-1800 (1983).

81. Allesandrini, A., Schmidt, E., Zilliken, F., and György, P. Enzymatic synthesis of β-D-galactosides of *N*-acetyl-D-glucosamine by various mammalian tissues, *J. Biol. Chem.* *220*, 71-78 (1956).

82. Zilliken, F., Smith, P. N., Rose, C. S., and György, P. Synthesis of 4-*O*-β-D-galactopyranosyl-*N*-acetyl-D-glucosamine by intact cells of *Lactobacillus bifidus, J. Biol. Chem.* *217*, 79-82 (1955).

83. Nilsson, K. G. I. Enzymic synthesis of di- and trisaccharide glycosides using glycosidases and β-D-galactoside 3-α-sialyltransferase, *Carbohydr. Res. 188*, 9-17 (1989).

84. Dahmén, J., Frejd, T., Lave, T., Lindh, F., Magnusson, G., Noori, G., and Pålsson, K. 4-*O*-α-D-Galactopyranosyl-D-galactose: Efficient synthetic routes from polygalacturonic acid, *Carbohydr. Res. 113*, 219-224 (1983).

85. Kobata, A., and Takasaki, S. *endo-β*-Galactosidase and *endo-α-N*-acetyl-galactosaminidase from *Diplococcus pneumoniae, Methods Enzymol. 50*, 560-567 (1978).

86. Hedbys, L., Johansson, E., Mosbach, K., Larsson, P.-O., Gunnarsson, A., and Svensson, S. Synthesis of 2-acetamido-2-deoxy-3-*O*-(β-D-galacto-pyrano-

syl)-D-galactose by the sequential use of β-galactosidases from bovine testes and *E. coli, Carbohydr. Res. 186*, 217-223 (1989).

87. Rabate, J. Contribution à l'étude biochemique des salicacées, *Bull. Soc. Chim. Biol. 17*, 572-601 (1935).

88. Nilsson, K. G. I. A simple strategy for changing the regioselectivity of glycosidase-catalyzed formation of disaccharides, *Carbohydr. Res. 167*, 95-103 (1987).

89. Nilsson, K. G. I. Glycosidase- and sialyltransferase-catalyzed synthesis of sialic acid containing trisaccharides, in *Sialic Acids 1988* (R. Schauer and T. Yamakawa, eds.), Kieler Verlag, Kiel, 1988, pp. 28-29.

90. Nilsson, K. G. I. A simple strategy for changing the regioselectivity of glycosidase-catalyzed formation of disaccharides II. Enzymic synthesis in situ of various acceptor glycosides, *Carbohydr. Res. 180*, 53-59 (1988).

91. Nilsson, K. G. I. Influence of various parameters on the yield and regioselectivity of glycosidase-catalyzed formation of oligosaccharide glycosides, *Ann. N.Y. Acad. Sci. 542*, 383-389 (1988).

92. Nilsson, K. G. I. Enzymic synthesis of HexNAc-containing disaccharide glycosides, *Carbohydr. Res. 204*, 79-83 (1990)..

93. Wallenfels, K., and Malhotra, O. P. Galactosidases, *Adv. Carbohydr. Chem. Biol. 16*, 239-298 (1961).

94. Huber, R. E. Kurz, G., and Wallenfels, K. A quantitation of the factors which affect the hydrolase and transgalactosylase activities of β-galactosidase (*E. coli*) on lactose, *Biochemistry, 15*, 1994-2001 (1976).

95. Nilsson, K. G. I. A comparison of the enzyme catalysed synthesis of peptides and oligosaccharides in various hydroorganic solutions using the nonequilibrium approach, in *Studies in Organic Chemistry*, Vol. 29 (C. Laane, J. Tramper, and M. D. Lilly, eds.), Elsevier, New York, 1987, pp. 369-374.

96. Hedbys, L., Larsson, P.-O. Mosbach, K., Svensson, S. Synthesis of the disaccharide 6-*O*-β-D-galactopyranosyl-2-acetamido-2-deoxy-D-galactose using immobilised β-galactosidase, *Biochem. Biophys. Res. Commun. 123*, 8-15 (1984).

97. Kuhn, R., Baer, H. H., and Gauhe, A. Chemische und enzymatische Synthese des 6-β-D-galactosido-*N*-acetyl-D-glucosamine, *Chem. Ber. 188*, 1713-1723 (1955).

98. Reglero, A., and Cabezas, J. A. Glycosidases of molluscs: Purification and properties of α-L-fucosidase from *Chamelea gallina* L., *Eur. J. Biochem. 66*, 379-387 (1976).

99. Huber, R. E., and Hurlburt, K. L. Reversion reactions of β-galactosidase (*E. coli*), *Arch. Biochem. Biophys. 246*, 411-418 (1986).

100. Elices, M. J., and Goldstein, I. W. Biosynthesis of bi-, tri-, and tetraantennary oligosaccharides containing α-D-galactosyl residues at their nonreducing termini, *J. Biol. Chem. 264*, 1375-1380 (1989).

101. Nilsson, K. G. I. Enzymatic synthesis of trisaccharides, in *Proceedings of the Vth European Symposium on Carbohydrates,* Prague, August 21-25, 1989, p. C-76.

102. Trimble, R. B., Atkinson, P. H., Tarentino, A. L. Plummer, T., Maley, F., and

Tomer, K. B. Transfer of glycerol by *endo-β-N*-acetylglucosaminidase F to oligosaccharides during chitobiose core cleavage, *J. Biol. Chem. 261,* 12000–12005 (1986).

103. Bardales, R. M., and Bhavanandan, V. P. Transglycosylation and transfer reaction activities of *endo-α-N*-acetyl-D-galactosaminidase from *Diplococcus (Streptococcus) pneumoniae, J. Biol. Chem. 264,* 19893–19897 (1989).

104. Nilsson, K. G. I. A method of controlling the regioselectivity of glycosidic bonds, European Patent Application, 86850414.3 (1986).

105. Bieg, T., and Szeja, W. Isomerization and cleavage of allyl ethers of carbohydrates by *trans*-(Pd(NH₃)₂Cl₂), *J. Carbohydr. Chem. 4,* 441–446 (1985).

106. Takeo, K., Kitajima, M., and Fukatsu, T. The Koenigs-Knorr reaction of benzyl 4,6-*O*-benzylidene-β-D-galactopyranoside with 2,3,4,6-tetra-*O*-acetyl-α-D-galactopyranosyl bromide, *Carbohydr. Res. 112,* 158–164 (1983).

107. Lipshutz, B. H., Pegram, J. J., and Morey, M. C. Chemistry of β-trimethylsilylethanol. II. A new method for protection of an anomeric center in pyranosides, *Tetrahedr. Lett. 22,* 4603–4606 (1981).

108. Fugedi, P., Garegg, P. J., Lönn, H., and Norberg, T. *Glycoconjugate J. 4,* 97–108 (1987).

109. Therisod, M., and Klibanov, A. M. Facile enzymatic preparation of monoacylated sugars in pyridine, *J. Am. Chem. Soc. 108,* 5638–5640 (1986).

110. Kloosterman, M., de Nijs, M. P., Weinen, J. G. J., Schoemaker, H. E., and Meijer, E. M. Regioselective hydrolysis of carbohydrate secondary acyl esters by lipases, *J. Carbohydr. Chem. 8,* 333–341 (1989).

111. Usui, T., Hayashi, Y., Nanjo, F., and Ishido, Y. Enzymatic synthesis of *p*-nitrophenyl *N,N,N,N,N*-pentaacetyl-β-chitopentaoside in water-methanol system; significance as a substrate for lysozyme assay, *Biochem. Biophys. Acta, 953,* 179–184 (1988).

112. Mozaffar, Z., Nakanishi, K., and Matsuno, R. Continuous production of galacto-oligosaccharides from lactose using immobilized β-galactosidase from *Bacillus circulans, Appl. Microbiol. Biotechnol. 25,* 224–228 (1986).

113. Naoi, M., Kiuchi, K., Sato, T., Morita, M., Tosa, T., Chibata, I., and Yagi, K. Alteration of the substrate specificity of *Aspergillus oryzae* β-galactosidase by modification with polyethylene glycol, *J. Appl. Biochem. 6,* 91–102 (1984).

114. Wallenfels, K. Transgalaktosylierung, *Bull. Soc. Chim. Biol. 42,* 1715–1729 (1960).

115. Ajisaka, K., Nishida, H., and Fujimoto, H. Use of an activated carbon column for the synthesis of disaccharides by use of a reversed hydrolysis activity of β-galactosidase, *Biotechnol. Lett. 9,* 387–392 (1987).

116. Hill, A. C. Reversible zymohydrolysis, *J. Chem. Soc. 73,* 634–658 (1898).

117. Adachi, S., Ueda, Y., and Hashimoto, K. Kinetics of formation of maltose and isomaltose through condensation of glucose by glucoamylase, *Biotechnol. Bioeng. 26,* 121–127 (1984).

118. Johansson, E., Hedbys, L., Mosbach, K., Larsson, P.-O., Gunnarsson, A., and Svensson, S. Studies of the reversed α-mannosidase reaction in high concentration of mannose, *Enzyme Microb. Technol. 11,* 349–352 (1989).

119. Tewari, Y. B., and Goldberg, R. N. Thermodynamics of hydrolysis of disaccharides, *J. Biol. Chem. 264*, 3966-3971 (1989).

120. Ajisaka, K., Nishida, H., and Fujimoto, H. The synthesis of oligosaccharides by the reversed hydrolysis reaction of β-glucosidase at high substrate concentration and at high temperature, *Biotechnol. Lett. 9*, 243-248 (1987).

121. Back, J. F., Oakenfull, D., and Smith, M. B. Increased thermal stability of proteins in the presence of sugars and polyols, *Biochemistry, 18*, 5191 (1979).

122. Johansson, E., Hedbys, L., Mosbach, K. Larsson, P.-O., Gunnarsson, A., and Svensson, S. Glycosidases in carbohydrate synthesis, *Proceedings of the XIVth International Carbohydrate Symposium*, Stockholm, 1988, p. B125.

123. Cori, C. F., Schmidt, G., and Cori, G. T. The synthesis of a polysaccharide from glucose-1-phosphate in muscle extract, *Science, 89*, 464-465 (1939).

124. Beyer, T. A., Sadler, J. E., Rearick, J. I., Paulson, J. C., and Hill, R. L. Glycosyltransferases and their use in assessing oligosaccharide structure and structure-function relationships, *Adv. Enzymol. 52*, 23-176 (1981).

125. Brockhausen, I., Hull, E., Hindsgaul, O., Schachter, H., Shah, R. N., Michnick, S. W., and Carver, J. P. Control of glycoprotein synthesis, *J. Biol. Chem. 264*, 11211-11221 (1989).

126. Paulson, J. C., and Colley, K. J. Glycosyltransferases, structure, localisation and control of cell-type specific glycosylation, *J. Biol. Chem. 264*, 17615-17618 (1989).

127. Sadler, J. E., Beyer, T. A., Oppenheimer, C. L., Paulson, J. C., Prieels, J.-P., Rearick, J. I., and Hill, R. L. Purification of mammalian glycosyltransferases, *Methods Enzymol. 83*, 458-514 (1982).

128. Whitehead, J. S., Bella, A., and Kim, Y. S. An *N*-acetylgalactosaminyltransferase from human blood group A plasma, *J. Biol. Chem. 249*, 3442-3447 (1974).

129. Weinstein, J., Souza-e-Silva, U., and Paulson, J. C. Purification of Galβ1-4GlcNAc α2-6-sialyltransferase and a Galβ1-3(4)GlcNAc α2-3-sialyltransferase to homogeneity from rat liver, *J. Biol. Chem. 257*, 13835-13844 (1982).

130. Barker, R., Olsen, K. W., Shaper, J. H., and Hill, R. L. Agarose derivatives of uridine phosphate and *N*-acetylglucosamine for the purification of a galactosyltransferase, *J. Biol. Chem. 247*, 7135-7147 (1972).

131. Palcic, M. M., Venot, A. P., Ratcliffe, R. M., and Hindsgaul, O. Enzymic synthesis of oligosaccharides terminating in the tumor-associated sialyl-Lewis-a determinant, *Carbohydr. Res. 190*, 1-11 (1989).

132. Nunez, H. A., and Barker, R. Enzymatic synthesis and carbon-13 nuclear magnetic resonance conformational studies of disaccharides containing β-D-galactopyranosyl and β-D-(1-^{13}C)galactopyranosyl residues, *Biochemistry, 19*, 489-495 (1980).

133. Rosevear, P. R., Nunez, H. A., and Barker, R. Synthesis and solution conformation of the type 2 blood group oligosaccharide Fucα1-2Galβ1-4GlcNAc, *Biochemistry, 21*, 1421-1431 (1982).

134. Gillies, D. R. B., Resnick, D., and Glick, M. C. Affinity purification of GDP-L-

Fuc: *N*-acetyl-β-D-glucosaminide α1-3-fucosyltransferase from human neuroblastoma, in *Proceedings of the Xth International Symposium on Glycoconjugates*, Jerusalem, 1989, pp. 216-217.

135. Takasaki, S., and Kobata, A. Chemical characterisation and distribution of ABO blood group active glycoprotein in human erythrocyte membrane, *J. Biol. Chem.* *251*, 3610-3615 (1976).

136. Rajan, V. P., Larsen, R. D., Ajmera, S., Ernst, L. K., and Lowe, J. B. A cloned human DNA restriction fragment determines expression of a GDP-L-fucose: β-D-galactoside 2-α-L-fucosyltransferase in transfected cells, *J. Biol. Chem.* *264*, 11158-11167 (1989).

137. Joziasse, D. H., Shaper, J. H., van den Eijnden, D. H., van Tunen, A. J., and Shaper, N. L. Bovine α1-3galactosyltransferase: Isolation and characterisation of a cDNA clone, *J. Biol. Chem.* *264*, 14290-14297 (1989).

138. Paulson, J. C., Weinstein, J., and Schauer, A. Tissue-specific expression of sialyltransferases, *J. Biol. Chem.* *264*, 10931-10934 (1989).

139. Larsen, R. D., Rajan, V. P., Ruff, M. M., Kukowska-Latallo, J., Cummings, R. D., and Lowe, J. B. Isolation of cDNA encoding a murine UDP-galactose: β-D-galactosyl-1,4-*N*-acetyl-D-glucosaminide α1,3-galactosyltransferase: Expression cloning by gene transfer, *Proc. Natl. Acad. Sci. USA*, 8227-8231 (1989).

140. Colley, K. J., Lee, E. U., Adler, B., Browne, J. K., and Paulson, J. C. Conversion of a Golgi apparatus sialyltransferase to a secretory protein by replacement of the NH_2-terminal signal anchor with a signal peptide, *J. Biol. Chem.* *264*, 17619-17622 (1989).

141. Wong, S.-H., Haynie, S. L., and Whitesides, G. M. Enzyme-catalyzed synthesis of *N*-acetyllactosamine with in situ regeneration of uridine 5′-diphosphate glucose and uridine 5′-diphosphate galactose, *J. Org. Chem.* *47*, 5416-5418 (1982).

142. Kean, E. L., and Roseman, S. CMP-sialic acid synthetase, *Methods Enzymol.* *8*, 208-215 (1966).

143. Schauer, R., Wember, M., and Ferreira do Amaral, C. Synthesis of CMP-glycosides of radioactive *N*-acetyl-, *N*-glycoloyl-, *N*-acetyl-7-*O*-acetyl-, and *N*-acetyl-8-*O*-acetylneuraminic acids by CMP-sialate synthase from bovine submaxillary glands, *Hoppe-Seylers Z. Physiol. Chem.* *353*, 883-886 (1972).

144. Corfield, A. P., Schauer, R., and Wember, M. The preparation of CMP-sialic acids by using CMP-acylneuraminate synthase from frog liver immobilised on Sepharose 4B, *Biochem. J.* *177*, 1-7 (1979).

145. Higa, H. H., and Paulson, J. C. Sialylation of glycoprotein oligosaccharides with *N*-acetyl-, *N*-glycolyl-, and *N*-*O*-diacetylneurmainic acids, *J. Biol. Chem.* *260*, 8838-8849 (1985).

146. Nilsson, K. G. I., and Gudmundsson, B.-M. G. Synthesis of CMP-Neu5Ac and Neu5Acα2-3Galβ1-3GalNAcα-OEt with immobilised CMP-sialate synthase and β-galactoside α2-3-sialyltransferase, in *Proceedings: Sialic Acids 1988* (R. Schauer and T. Yamakawa, eds.) Kieler Verlag, Kiel, 1988, pp. 30-31.

147. Augé, C., and Gautheron, C. An efficient synthesis of cytidine monophophospho-sialic acids with four immobilized enzymes, *Tetrahedron Lett.* *29*, 789-790 (1988).

148. Simon, E. S., Bednarski, M. D., and Whitesides, G. M. Synthesis of CMP-Neu5Ac from *N*-acetylglucosamine: Generation of CTP from CMP using adenylate kinase, *J. Am. Chem. Soc. 110*, 7159-7163 (1988).

149. Ropp, P., and Cheng, P.-W. Enzymatic preparation of UDP-GlcNAc by a batch procedure, in *Proceedings of the Xth International Symposium on Glycoconjugates*, Jerusalem, 1989, p. 213.

150. Ginsburg, V. Sugar nucleotides and the synthesis of carbohydrates, *Adv. Enzymol. 26*, 35-85 (1964).

151. Nilsson, K. G. I., and Gudmundsson, B.-M. E. Synthesis of CMP-Neu5Ac and Neu5Acα2-3Galβ1-3GalNAcα-OEt with CMP-sialate synthase and β-D-galactoside α2-3-sialyltransferase immobilised to tresyl chloride activated agarose, *Methods Mol. Cell. Biol. 1*, 195-202 (1990).

152. Christian, R., Schreiner, E., Zbiral, E., and Schulz, G. The side-chain conformations of *N*-acetyl-7-, 8-, 9-deoxy-, and -4,7-dideoxy-neuraminic acid and their effect on the activation of CTP:*N*-acylneuraminic acid cytidylyltransferase, *Carbohydr. Res. 194*, 49-61 (1989).

153. Gross, H. J., Bünsch, A., Paulson, J. C., and Brossmer, R. Activation and transfer of novel synthetic 9-substituted sialic acids, *Eur. J. Biochem. 168*, 595-602 (1987).

154. Conradt, H. S., Bunsch, A., and Brossmer, R. Preparation of 9-fluoro-9-deoxy-*N*-(2-^{14}C)acetylneuraminic acid, *FEBS Lett. 170*, 295-300 (1984).

155. Yates, A. D., and Watkins, W. M. The biosynthesis of blood group B determinants by the blood group A gene-spcified α-*N*-acetyl-D-galactosaminyltransferase, *Biochem. Biophys. Res. Commun. 109*, 958-965 (1982).

156. Prieels, J.-P., and Beyer, T. A. Purification and properties of GDP-fucose: *N*-Acetylglucosaminide α1,3-fucosyltransferase, *Fed. Proc. 38*, 631 (1979).

157. Van Pelt, J., Dorland, L., Duran, M., Hokke, C. H., Kamerling, J. P., and Vliegenthart, J. F. G. Transfer of sialic acid in α2-6 linkage to mannose in Manβ1-4GlcNAcβ1-4GlcNAc by the action of Galβ1-4GlcNAc α2-6-sialyltransferase, *FEBS Lett. 256*, 179-184 (1989).

158. Srivastava, G., Alton, G., and Hindsgaul, O., *Carbohydr. Res. 207*, 259-276 (1990).

159. Weinstein, J., Souza-e-Silva, U., and Paulson, J. C. Sialylation of glycoprotein oligosaccharides *N*-linked to asparagine, *J. Biol. Chem. 257*, 13845-13853 (1982).

160. Shibaev, V. N. Biosynthesis of *Salmonella O*-antigenic polysaccharides: Specificity of glycosyltransferases, *Pure Appl. Chem. 50*, 1421-1436 (1978).

161. Morin, M. J., Porter, C. W., Petrie, C. R., Korytnyk, W., and Bernacki, R. J. *Biochem. Pharm. 32*, 553 (1983).

162. McDowell, W., Grier, T. J., Rasmussen, J. R., and Schwarz, R. T. The role of C-4-substituted mannose analogues in protein glycosylation, *Biochem. J. 248*, 523-531 (1987).

163. Gross, H. J., Sticher, U., and Brossmer, R. A highly sensitive fluorometric assay

for sialyltransferase activity using CMP-9-fluoresceinyl-NeuAc as donor, *Anal. Biochem, 186*, 127-134 (1990).

164. Schwyzer, M., and Hill, R. L. Porcine A blood group-specific *N*-acetylgalactosaminyltransferase, *J. Biol. Chem. 252*, 2338-2345 (1977).

165. Kobata, A., Grollman, E. F., and Ginsburg, V. An enzymatic basis for blood type B in humans, *Biochem. Biophys. Res. Commun. 32*, 272-277 (1968).

166. Nishikawa, Y., and Schachter, H. Purification and characterisation of rabbit liver UDP-*N*-acetyl-glucosamine: α3-mannoside β2-*N*-acetyl glucosaminyltransferase, in *Proceedings of the International Symposium on Glycoconjugates*, Lille, 1987, pp. E96.

167. Rearick, J. I., Sadler, J. E., Paulson, J. C., and Hill, R. L. Enzymatic characterisation of β-D-galactoside α2-3-sialyltransferase from procine submaxillary gland, *J. Biol. Chem. 254*, 4444-4451 (1979).

168. Johnson, P. H., and Watkins, W. M. Purification of α-3-L-fucosyltransferase from human liver, in *Proceedings of the Xth International Symposium on Glycoconjugates*, Jerusalem, 1989, pp. 214-215.

169. Trayer, I. P., and Hill, R. L. The purification and properties of the A protein of lactose synthetase, *J. Biol. Chem. 246*, 6666-6675 (1971).

170. Zehavi, U., Sadeh, S., and Herchman, M. Enzymic synthesis of oligosaccharides on a polymer support. Light-sensitive, substituted polyacrylamide beads, *Carbohydr. Res. 124*, 23-34 (1983).

171. Zehavi, U., and Herchman, M. Enzymic synthesis of oligosaccharides on an α-chymotrypsin-sensitive polymer. *O*-(β-D-galactopyranosyl)-(1-4)-*O*-(β-D-glucopyranosyl)-(1-4)-D-glycopyranose, *Carbohydr. Res. 133*, 339-342 (1984).

172. Brew, K., Vanaman, T. C., and Hill, R. L. The role of α-lactalbumin and the A protein in lactose synthetase: A unique mechanism for the control of a biological reaction, *Proc. Natl. Acad. Sci. USA, 59*, 491-497 (1968).

173. Brodbeck, U., Denton, W. L., Tanahashi, N., and Ebner, K. E. The isolation and identification of the B protein of lactose synthetase as α-lactalbumin, *J. Biol. Chem. 242*, 1391-1397 (1967).

174. Johnson, P. H., Watkins, W. M., and Donald, A. S. R. Further purification of the Le-gene associated α-L-fucosyltransferase from human milk, in *Proceedings of the IXth International Symposium on Glycoconjugates*, Lille, 1987, pp. E107.

175. Augé, C., David, S., Mathieu, C., and Gautheron, C. Synthesis with immobilised enzymes of two trisaccharides, one of them active as the determinant of a stage antigen, *Tetrahedron Lett. 25*, 1467-1470 (1984).

176. Augé, C., Mathieu, C., and Merienne, C. The use of an immobilised cyclic multienzyme system to synthesise branched penta- and hexasaccharides associated with blood-group I epitopes, *Carbohydr. Res. 151*, 147-156 (1986).

177. Sabesan, S., and Paulson, J. C. Combined chemical and enzymatic synthesis of sialyloligosaccharides and characterisation by 500 MHz ^{1}H and ^{13}C NMR spectroscopy, *J. Am. Chem. Soc. 180*, 2068-2080 (1986).

178. David, S., and Augé, C. Immobilised enzymes in preparative carbohydrate chemistry, *Pure Appl. Chem. 59*, 1501-1508 (1987).

179. de Heij, H. T., Kloosterman, M., Koppen, P. L., van den Boom, J. H., and van den Eijnden, D. H. Combined chemical and enzymatic synthesis of a disialylated tetrasaccharide analogous to M and N blood group determinants of glycophorin A, *J. Carbohydr. Chem. 7*, 209-222 (1988).

180. Palcic, M. M., Srivastava, O. P., and Hindsgaul, O. L. Transfer of D-galactosyl groups to 6-*O*-substituted 2-acetamido-2-deoxy-D-glucose residues by use of bovine D-galactosyltransferase, *Carbohydr. Res. 159*, 315-324 (1987).

181. Thiem, J., and Treder, W. F. Synthese des Trisaccharids Neu-5-Ac-α2-6Galβ1-4GlcNAc mit immobilisierten Enzymen, *Ang. Chem. 98*, 1100-1101 (1986).

182. Hindsgaul, O., Tahir, S. H., Srivastava, O. P., and Pierce, M. The trisaccharide β-D-GlcNAc-(1-2)-α-D-Man-(1-6)-β-D-Man, as its 8-methoxycarbonyloctyl glycoside, is an acceptor selective for *N*-acetylglucosaminyltransferase V, *Carbohydr. Res. 173*, 263-272 (1988).

183. Crawley, S. C., Hindsgaul, O., Ratcliffe, R. M., Lamontagne, L. R., and Palcic, M. M. A plant fucosyltransferase with human Lewis blood group specificity, *Carbohydr. Res. 193*, 249-256 (1989).

184. Augé, C., Gautheron, C., and Pora, H. Enzymic cynthesis of the sialylglycopeptide, α-D-Neu5Ac-2-6-β-D-Gal-1-4-β-D-GlcNAc-1-4N-L-Asn, *Carbohydr. Res. 193*, 288-293 (1989).

185. Wiemann, T., and Thiem, J. Combined chemoenzymatic synthesis of *N*-glycoprotein substructures, in *Proceedings of the International Symposium on Glycoconjugates*, Jerusalem, 1989, pp. 297-298.

186. Ring, M., Bader, D. E., and Huber, R. E. Site-directed mutagenesis of β-galactosidase (E. coli) reveals that Tyr-503 is essential for activity, *Biochem. Biophys. Res. Commun. 152*, 1050-1055 (1988).

187. Bader, D. E., Ring, M., and Huber, R. E. Site-directed mutagenic replacement of Glu-461 with Gln in β-galactosidase (E. coli): evidence that Glu-461 is important for activity, *Biochem. Biophys. Res. Commun. 153*, 301-305 (1988).

ABBREVIATIONS

Asn	Asparagine
Bn	Benzyl
Bz	Benzoyl
Cer	Ceramide
CMP	Cytidine 5′-monophosphate
CMP–Neu5Ac	Cytidine 5′-monophospho-N-*actylneuraminic acid*
CTP	Cytidine 5′-triphosphate
Et	Ethyl
Fuc	L-Fucopyranose
Gal	D-Galactopyranose
GalNAc	2-Acetamido-2-deoxy-D-galactopyranose
CDP	Guanosine 5′-diphosphate

GlcNAc	2–Acetamido–2–deoxy–D–glucopyranose
Man	D–Mannopyranose
Me	Methyl
Neu5Ac	*N*–Acetylneuraminic acid
Ph	Phenyl
Ser	Serine
Thr	Threonine
UDP	Uridine 5′–diphosphate

Abbreviated nomenclature of oligosaccharides according to IUB–IUPAC recommendations [*J. Biol. Chem.* *257*, 3347–3351, (1982)].

5

Enzymatic Modification of Steroids

Sergio Riva

Istituto di Chimica degli Ormoni, Consiglio Nazionale delle Ricerche, Milan, Italy

INTRODUCTION

The structural and functional elaboration of the steroid skeleton has been a fertile field for organic chemists since the appearance of the first elucidations of the structure of molecules belonging to this important class of compounds. Moreover, the pharmaceutical properties and the high cost of manufacturing steroids made any approach that permits mild and efficient modifications of these compounds a noteworthy achievement. As a result of this effort, highly selective synthetic procedures for the transformation of complex derivatives have been developed [1–5].

This kind of biologically active molecule, containing different sensitive functional groups, appeared to be more and more suitable as a vehicle for exploiting the potentialities of enzymes in organic synthesis, mainly consisting of mild reaction conditions and excellent regio- and/or stereoselectivity. The literature of this field has been spread widely in chemical and biological journals, making available an enormous amount of scientific data.

Microbial transformations have been discussed in specific books [6,7]. Therefore this chapter is limited to the applications of isolated enzymes, crude or purified. At first glimpse, it is evident that there are two main parameters involved in any enzymatic transformation:

179

The origin and catalytic properties of the different enzymes
The milieus in which the reactions have been performed

The latter parameter has been investigated with particular thoroughness in the past few years. The same enzymes have been employed in different environments ranging from "natural" aqueous buffers to pure anhydrous organic solvents. Quite interestingly, enzymes not only maintained their catalytic action in these unconventional milieus, but also showed new and sometimes unexpected properties.

Taking into account this modern perspective of enzymatic catalysts, this chapter is divided into the following sections:

Reactions in aqueous solutions
Reactions in monophasic solutions containing water-miscible solvents
Reactions in biphasic systems
Reactions in reversed micelles
Reactions in organic solvents
Reactions in supercritical carbon dioxide

A small section has also been dedicated to the analytical applications of such enzymantic systems.

ENZYMATIC REACTIONS IN AQUEOUS SOLUTIONS

Neutral steroids are compounds whose solubility in water is markedly low, ranging, with few exceptions, from 10^{-4} to 10^{-5} mole/L. This implies that steroid transformations, performed either by isolated enzymes or by fermentation processes, must be carried out using large reaction volumes. This is not the case with bile acid derivatives, which have a carboxylic group on the side chain (e.g., cholic acid, **1**), with their conjugates with glycine (e.g., glycocholic acid, **2**) and taurine (e.g., taurocholic acid, **3**) and with sulfuric or glucurinic acid

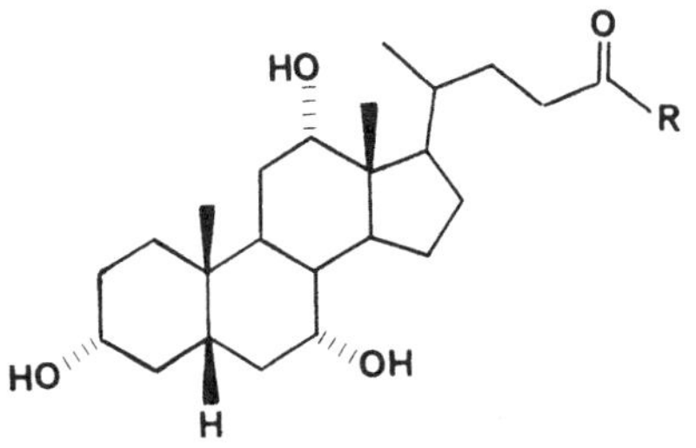

R = OH $\underline{1}$

R = NHCH$_2$COOH $\underline{2}$

R = NHCH$_2$SO$_3$H $\underline{3}$

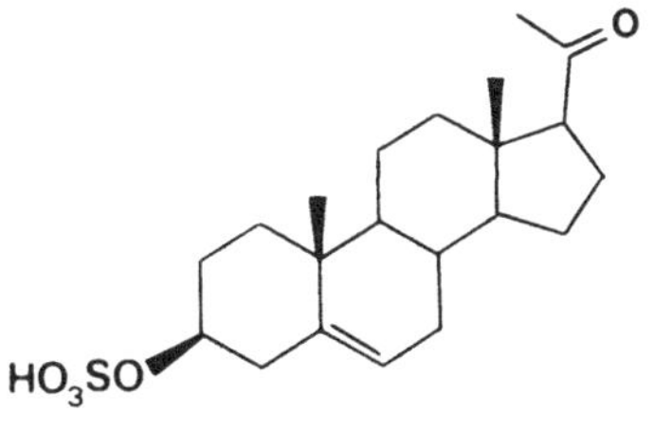

esters of neutral steroids (e.g. pregnenolone sulfate, **4**). Enzymatic transformations of these compounds in water have been reported, mainly involving hydrolases and dehydrogenases. On the other hand, transferases (e.g., glucosyltransferase and sulfotransferase) have been employed to synthesize water-soluble derivatives of neutral steroids.

Hydrolases

Sterol Sulfatase

Sterol sulfatases (sterol-sulfate sulfohydrolase) were reviewed in an early paper by Sandberg and Jenkins[8]. These enzymes have been isolated from molluscan (*Patella vulgata, Helix pomatia, Otala punctata,* etc.) and mammalian sources (liver, testis, placenta). Available data suggested that both the mammalian and molluscan enzymes are quite specific for 3β-sulfates, catalyzing the irreversible hydrolysis of the 5α and Δ^5 series of steroids. Little or no activity has been observed toward the other isomeric 3-sulfates, or toward 16α-, 17α-, 17β-, and 20α-, and 20β-sulfoxysteroids. Table 5.1 shows the relative rates of hydrolysis of certain Δ^5-3β-sulfoxysteroids by a microsomal cell fraction of human term placenta.

Table 5.1   Relative Rates of Hydrolysis of Δ^5-3β-Sulfoxy Compounds by a Microsomal Cell Fraction of Human Term Placenta [8]

Steroid	Relative rate
5-Androsten-3β,17β-diol-3-sulfate	20
5-Androsten-3β-17-oxo-3-sulfate	40
5-Pregnen-3β-ol-20-oxo-3-sulfate	70
5-Pregnen-3β,17α-diol-20-oxo-3-sulfate	100
Cholesterol sulfate	50

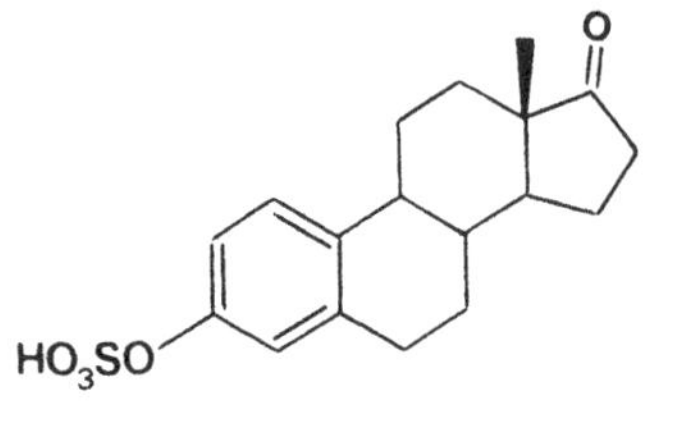

$\underline{5}$

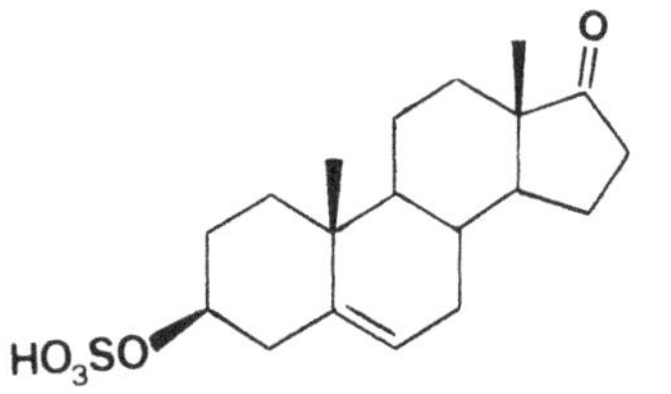

$\underline{6}$

Furthermore a different sulfatase (arysulfatase or arylsulfate sulfohydrolases) capable of hydrolyzing estrone sulfate (**5**) but not dehydroisoandrosterone sulfate (**6**) has been found in preparations of the fungus *Aspergillus oryzae* [8].

β-Glucoronidase

β-Glucoronidases from a number of sources—mammalian, molluscan, and microbial—are available commercially. These enzymes have been used in the characterization of natural products and in the preparation of samples prior to the estimation of urinary steroids [9].

Dehydrogenase

Hydroxysteroid dehydrogenases (HSDH) are NAD(P)-dependent oxidoreductases that catalyze the reversible oxidoreduction of the hydroxyl–keto groups of steroids with high regio- and stereospecificity. The development of effective methods for nicotinamide cofactor regeneration [10] and enzyme immobilization [11] has increased the potentialities of these catalysts for the synthesis of standard compounds, enzyme substrates, metabolites, and pharmaceuticals. The usefulness of the HSDHs for bile acids synthesis in phosphate buffer solution have been investigated with cholic acid (**1**) and dehydrocholic acid (**7**) as model substrates [12].

Cholic acid was oxidized regiospecifically on a preparative scale at each of the three possible hydroxyl groups (namely, $3\alpha, 7\alpha, 12\alpha$) to furnish the corresponding keto derivatives **8–10** (Fig. 5.1).

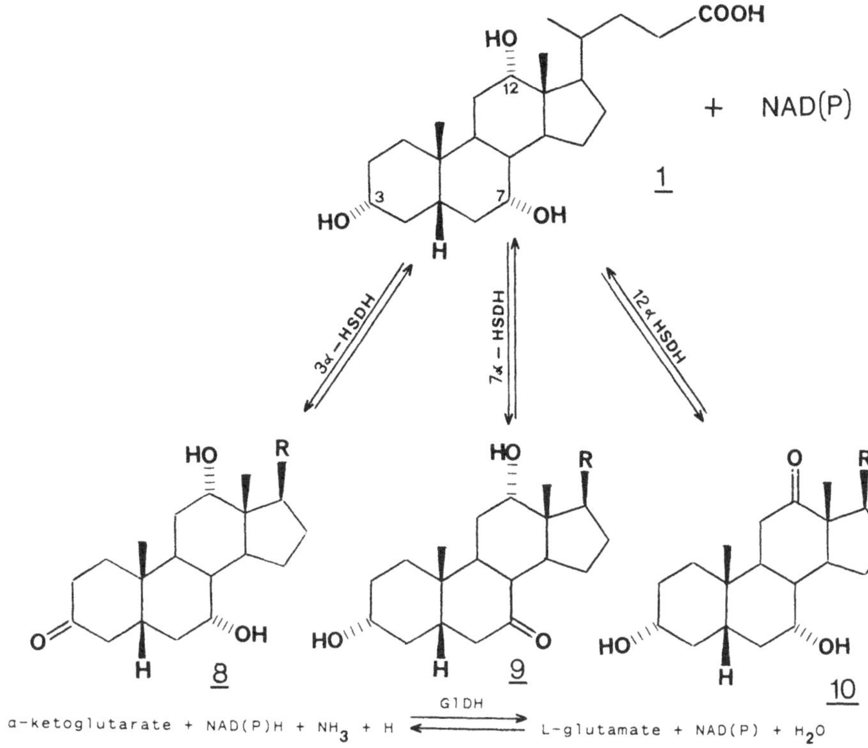

Figure 1 Regiospecific enzymatic oxidation of cholic acid (**1**) [12].

Dehydrocholic acid (**7**) was reduced regiospecifically and stereospecifi-cally at each of the three positions, thus giving 3α (**11**), 3β (**12**), 7α (**13**), 7β (**14**), or 12α (**15**) hydroxy derivatives (Fig. 5.2). The compounds were trans-formed quantitatively, and the products were 97–99% pure. The nicotinamide cofactors were regenerated enzymatically in situ. The enzymes were employed in the free form or they were immobilized on Sepharose CL-4B activated with tresyl chloride.

As a more general approach, the potential of these enzymatic processes to perform a specific chemical transformation, namely the hydroxyl α/β inver-sion at position 3 of the steroid skeleton, was also investigated [13]. This stereospcific transformation was carried out with numerous bile acids, differ-ing from each other in both the position of the hydroxyl groups and the length of the side chain (Fig. 5.3). Inversion was obtained in two steps through the

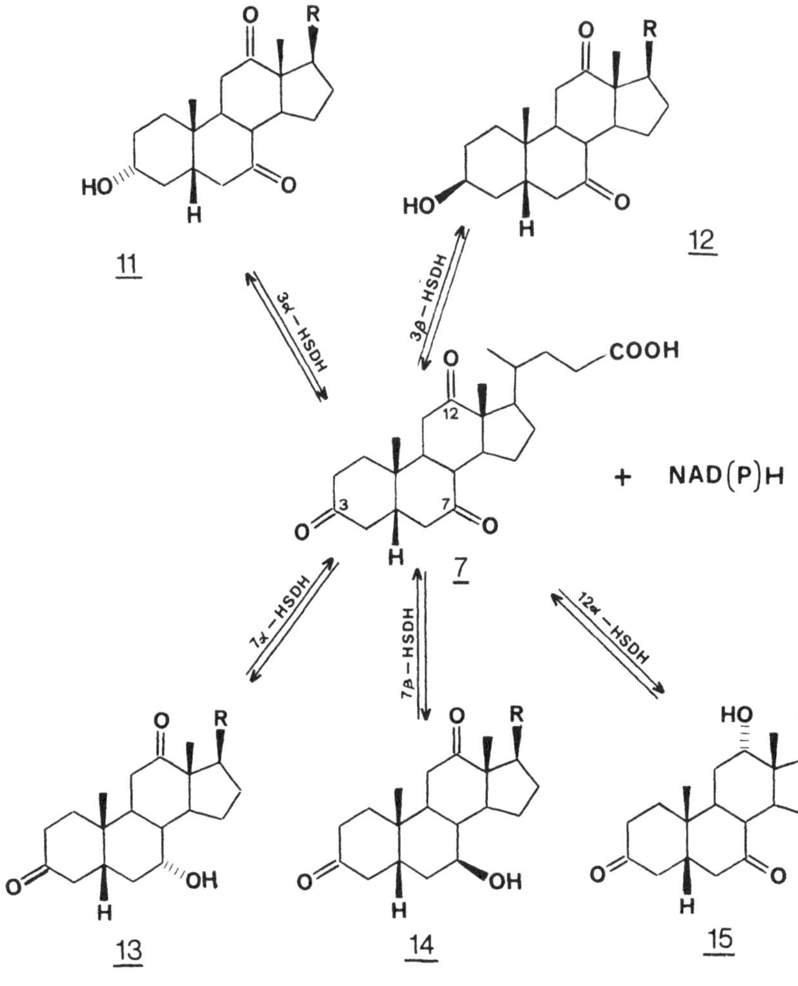

$$HCOOH + NAD \xrightleftharpoons{FDH} CO_2 + NADH + H$$

$$\text{or} \quad \text{D-Glucose} + NAD(P) \xrightleftharpoons{GlDH} \text{D-glucono-}\delta\text{-lactone} + NAD(P)H + H$$

Figure 2 Regio- and stereospecific reduction of dehydrocholic acid (7) [12].

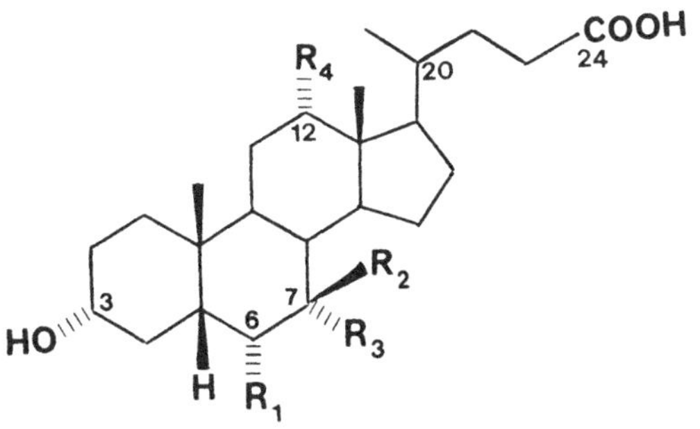

COMPOUND		R_1	R_2	R_3	R_4
Cholic acid	1	H	H	OH	OH
Iocholic acid	16	OH	H	OH	H
Chenodeoxycholic acid	17	H	H	OH	H
Ursodeoxycholic acid	18	H	OH	H	H
Deoxycholic acid	19	H	H	H	OH

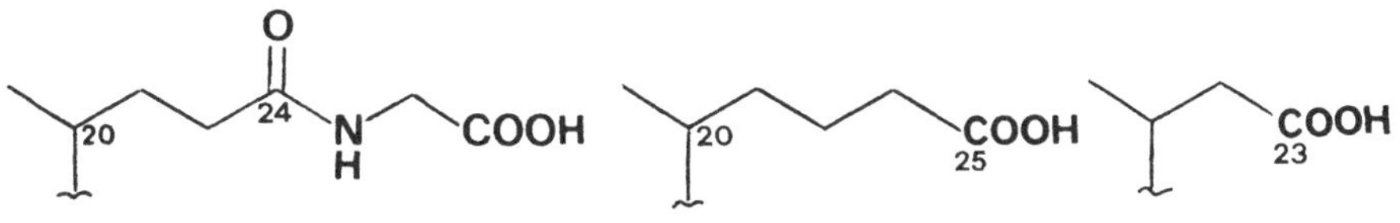

1a 1b 19a

Figure 3 3α-Hydroxy bile acids enzymatically transformed to the corresponding 3β-hydroxy derivatives [13].

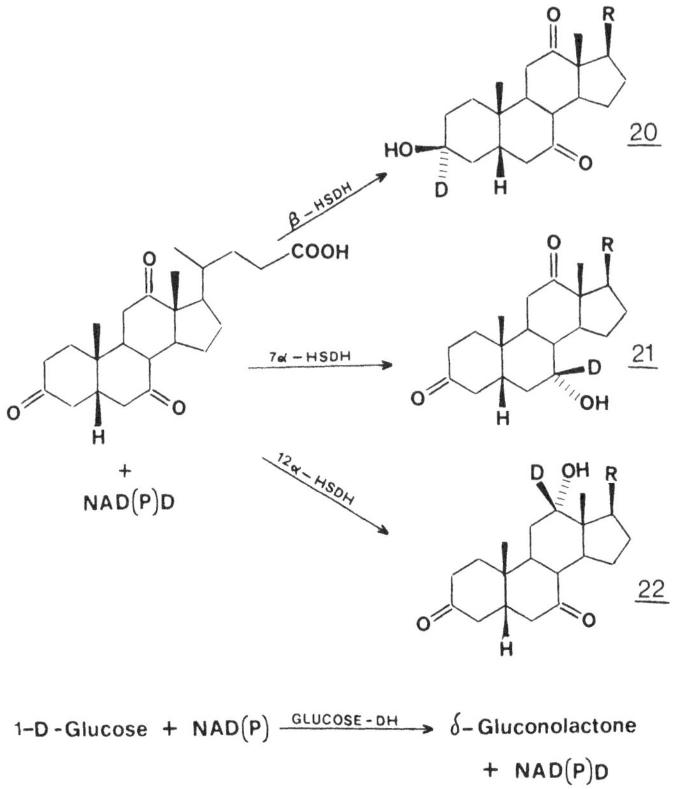

Figure 4   Synthesis of regio- and stereospecifically deuterated bile acids [13].

sequential use of the commercial enzymes 3α- and 3β-HSDH; these enzymes were employed in the free form of immobilized on Eupergit C. The transformations were practically quantitative and the products were more than 98% pure. NAD^+ was regenerated in situ with the pyruvate-lactic dehydrogenase system, and NADH was regenerated with the formate-formate dehydrogenase system.

Furthermore, the regio- and stereospecific syntheses of the deuteriated steroids **20–22** have been carried out on a preparative scale by coupling the reduction of dehydrocholic acid (**7**), catalyzed by a specific HSDH, to the oxidation of [1-^{2}H]glucose, catalyzed by glucose dehydrogenase (Fig. 5.4) [14]. The transfer of deuterium from glucose to bile acid was mediated by catalytic amounts of NAD(P) continuously recycled in situ. The isotopic purity of the deuteriated bile acids **20–22**, determined by 1H-NMR spectrometry, exceeded 94%.

One of the reactions described above, the synthesis of 12-ketochenodeoxycholic acid (**10**), has been studied in more detail [15-17], because of the interest of the pharmaceutical industry in this steroid. Compound **10** is in fact an essential intermediate in the synthesis of chenodeoxycholic acid (**17**) and ursodeoxycholic acid (**18**), which are used therapeutically in the dissolution of gallstones. Figure 5.5 compares the enzymatic and chemical syntheses of **10** from **1**. High concentrations (up to 4% w/v) of cholic acid could be completely and selectively converted to the desired product in one step, whereas with the chemical method [18] the yield was lower (<50%) and the product was contaminated with substances that give rise, when **10** is chemically reduced to **17**, to the toxic lithocholic acid.

To optimize the process, the properties of the enzymes 12α-HSDH (from *Clostridium*) and glutamate dehydrogenase (GlDH from *Proteus*), immobilized onto Sepharose CL-4B, have been studied [17].

Coenzyme recycling by free 12α-HDSH and GlDH, coimmobilized 12α-HSDH and GlDH, and separately immobilized 12α-HSDH and GlDH have been investigated following the transformation of **1** to **10**. The coimmobilized enzymes transformed the substrate with a rate very close to that of the free enzymes at all concentrations of coenzyme investigated [15-350μM; K_M(NADP), 15 μM for free 12α-HSDH, K_M(NADPH), 14μM for free GlDH], whereas with the separately immobilized enzymes the rate was markedly lower. The stability of the coimmobilized enzymes, tested under working conditions, was very high. Both enzymes maintained 70% of the initial activity after 2 months of continuous use. Furthermore, an efficient and cheap bioprocess for the production and purification of the noncommercial 12α-HSDH has been developed [19]. Studies are currently in progress, financed by an EEC research program, to develop a pilot reactor producing **10** on a multikilogram scale [20].

Transferase

Glucosyl Transferase

The transfer of glucose to the 17α-hydroxyl group of 17α-estradiol-3-glucuronoside has been shown in the microsomes of kidney and large intestine, as well as in those of the liver. Steroid glucosyl-, glucuronyl-, and N-acetylglucosaminyl transferases from rabbit liver microsomes have been partially purified [21].

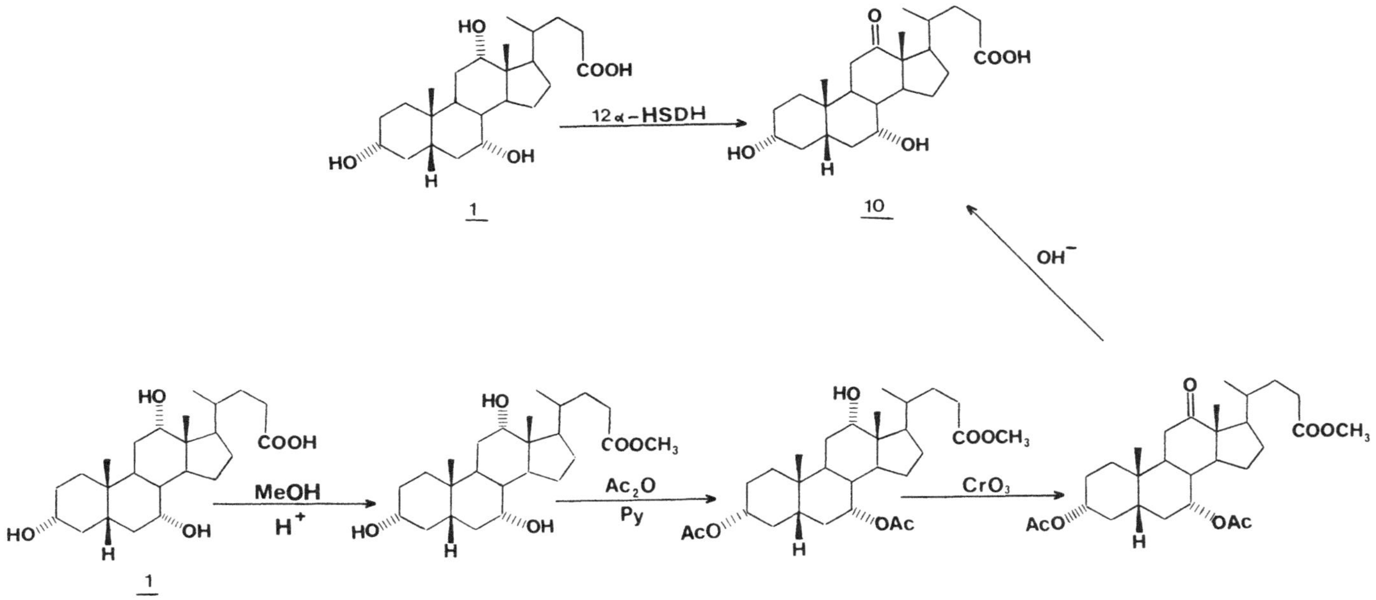

Figure 5   Comparison of enzymatic and chemical syntheses of 12-ketochenodeoxycholic acid (10) from cholic acid (1) [17,18].

The liver glucosyltransferase had a specific requirement for UDP-glucose and transferred glucose from this nucleotide to estrone, 17α-estradiol, and 17β-estradiol at the phenolic 3-hydroxyl group.

Sulfotransferase

Several papers have been published by Adams and coworkers concerning the isolation and characterization of sulfotransferases from different sources [22, and references therein].

Estrogen sulfotransferase, isolated from bovine adrenal glands or bovine placental tissue, sulfurylated the phenolic group of natural and synthetic estrogens, while other steroidic hydroxyl groups were unaffected.

On the other hand, hydroxysteroid sulfotransferase from human adrenal gland was able to catalyze sulfurylation on a wide range of C-18, C-19, and C-21 steroids to produce monosulfurylated derivatives. A parallel behavior was observed with a hydroxysteroid sulfotransferase isolated from rat liver.

ENZYMATIC REACTIONS IN MONOPHASIC SOLUTIONS CONTAINING WATER–MISCIBLE SOLVENTS

Addition of water-miscible organic solvents—such as acetone, ethanol, methanol, acetonitrile, and dioxane— has been the most obvious trick utilized to increase steroid solubility. However, while low concentrations of organic solvents affect enzyme stability and activity hardly at all (sometimes even increasing their stability or catalytic efficiency), higher solvent concentrations progressively give rise to inhibition, decreased specificity, and unfolding. Furthermore the use of a soluble cosolvent does not solve the problem of enzyme inhibition exerted by high substrate or product concentrations. For these reasons, the addition of soluble cosolvents usually has been limited to a few percent of the volume of the aqueous phase. Several examples reported in the literature, ranging from hydrolases to chloroperoxidases, from oxygenases to glycosidases, are discussed in this section.

Hydrolase

Hydrolytic cleavage of steroid esters has been observed in both microorganisms and animal tissues. Partially purified steroid esterases as well as other nonspecific hydrolases have been employed in buffer solutions containing dimethylformamide (DMF), methanol, or propylene glycol in a volume ratio ranging from 0.2 to 15%.

In a first report [23], enzymatic hydrolysis of steroidal acetates was achieved employing a commercially available α-amylase (named "diastase" in Ref. 23). This α-amylase showed an interesting selectivity in the hydrolysis. While the acetoxyl groups at positions C-16, C-17, C-20, and C-21 were hydrolyzed to the free alcohol, the other acetate were unaffected. For example, 11α,21-diacetoxy-17α-hydroxy-4-pregnene-3,20-dione (**23**), 15α,21-diacetoxy-17α-hydroxy-4-pregnene-3,20-dione (**24**), and 17α-hydroxy-19,21-diacetoxy-4-pregnene-3,20-(**25**), were hydrolyzed to the 11α-monoacetate (**23a**), 15α-monoacetate (**24a**) and 19-monoacetate (**25a**) in substantially quantitative yields.

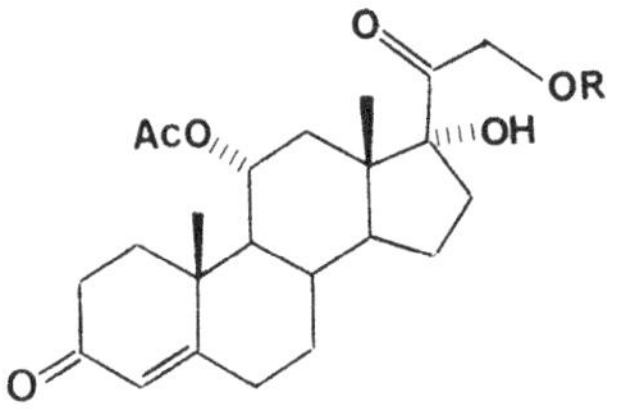

R= Ac , 23
R=H , 23a

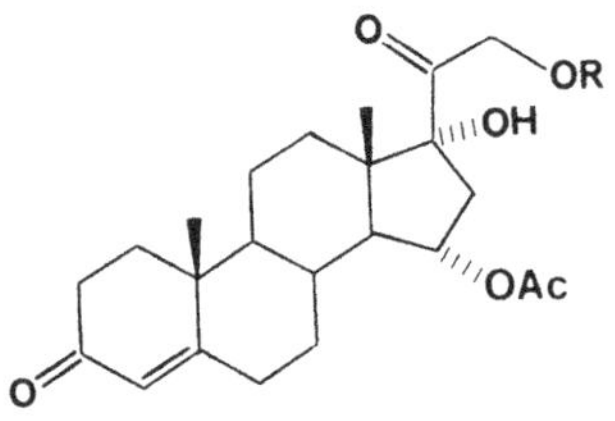

R=Ac , 24
R=H , 24a

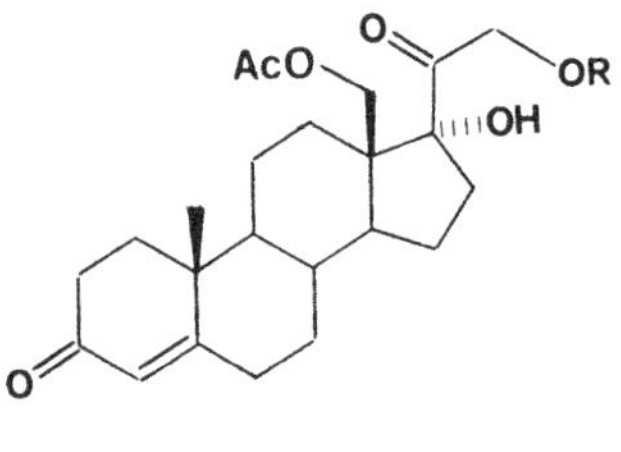

R=Ac , 25
R=H , 25a

In a later report [24], pig pancreatic amylase gave the most satisfactory results in the hydrolysis of 3β,16β-diacetoxy-5androsten-17-one (**26**), and 3β, 16β-dihydroxy-5-androsten-17-one (**26a**) was obtained in 64% yield after crystallization. It should be stressed here that compound **26a** was obtained in very low yield (<30%) with the classical chemical deacylation procedures, because of the strong tendency for 16β-hydroxy-17-oxoketols to isomerize to 17β-hydroxy-16-oxo compounds in the presence of acid or alkali.

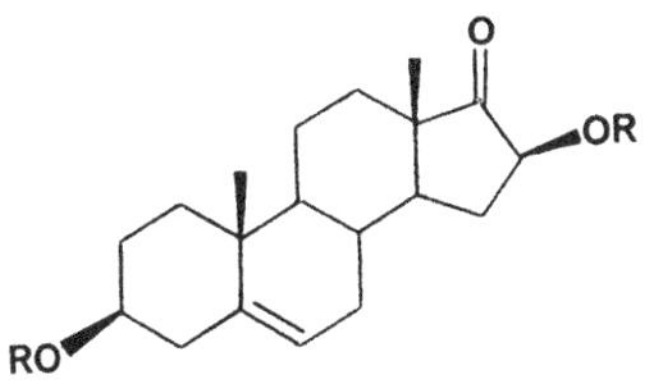

R= Ac , 26
R=H , 26a

In a review article [25], Rahim and Sih reported the purification of two bacterial steroid esterases, together with their properties and substrate specificities. The first enzyme, steroid esterase from *Nocardia restrictus*, did not hydrolyze succinate, benzoate, or propionate esters of testosterone (**27**), whereas acetate and pyruvate esters were readily cleaved. The substrate specificity and the relative rates of hydrolysis are shown in Table 5.2.

Table 5.2 Substrate Specificity of Purified Steroid Esterase from *Nocardia restrictus* [25]

Steroid	Relative hydrolysis
17α-Acetoxy-3-oxo-4-androstene	100
17β-Acetoxy-3-oxo-4-androstene	85
17α-Acetoxy-3,20-dioxo-4-pregnene	0
16α-Acetoxy-3,20-dioxo-4-pregnene	100
15α-Acetoxy-3,20-dioxo-4-pregnene	4
15β-Acetoxy-3,20-dioxo-4-pregnene	0
12α-Acetoxy-3,20-dioxo-5β-pregnene	0
11α-Acetoxy-3,20-dioxo-4-pregnene	0
11β-Acetoxy-3,17-dioxo-4-androstene	0
7α-Acetoxy-3,17-dioxo-4-androstene	0
7β-Acetoxy-3,17-dioxo-4-androstene	0
6α-Acetoxy-3,17-dioxo-1,4-androstadiene	50
6β-Acetoxy-3,17-dioxo-1,4-androstadiene	0
3β-Acetoxy-20-oxo-5α-pregnane	85
3α-Acetoxy-17-oxo-5α-androstane	80
3β-Acetoxy-20-oxo-5β-pregnane	0
3α-Acetoxy-20-oxo-5β-pregnane	0
3β-Acetoxy-20-oxo-5-pregnene	85

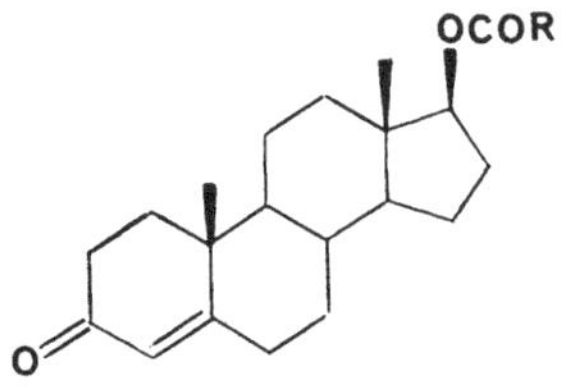

27

In a similar manner the second enzyme, steroid esterase from *Cylindrocarpon radicicola*, readily cleaved acetate, propionate, and pyruvate esters of testosterone (**27**), whereas testosterone succinate and benzoate were not affected. Acetoxy esters at positions 17α, 17β, 21, and 11β were hydrolyzed rapidly. On the other hand, acetoxy esters at positions 16α, 6α, 6β, 1α, 2α, and 3-(5α- or pregn-5-ene series) were hydrolyzed very slowly. The relative rates of hydrolysis of various steroid esters are listed in Table 5.3.

Chloroperoxydase

Chloroperoxydase catalyzes the peroxidative formation of the carbon-halogen bond according to Equation (1):

$$H_2O_2 + X^- + HA \longrightarrow AX + HO^- + H_2O$$

Table 5.3 Substrate Specificity of the Purified Steroid Esterase from *Cylindrocarpon radicicola* [25]

Steroid	Relative hydrolysis
17α-Acetoxy-3-oxo-4-androstene	100
17β-Acetoxy-3-oxo-4-androstene	78
17β-Pyruvoxy-3-oxo-4-androstene	55
17β-Propioxy-3-oxo-4-androstene	35
17β-Succinoxy-3-oxo-4-androstene	0
17β-Benzoyloxy-3-oxo-4-androstene	0
17α-Acetoxy-3,20-dioxo-4-pregnene	0
15α-Acetoxy-3,20-dioxo-4-pregnene	0
15β-Acetoxy-3,20-dioxo-4-pregnene	0
12α-Acetoxy-3,20-dioxo-5β-pregnane	0
11α-Acetoxy-3,20-dioxo-4-pregnene	0

where X- represents an oxidizable halogen anion (chloride, bromide, or iodide) and HA represents an acceptor molecule. A wide diversity of compounds, including β-keto acids, cyclic β-diketones, phenols and related aromatic compounds, and sulfides, can be transformed by this enzyme.

More specifically, a haloperoxydase isolated from the fungus *Caldaryomices fumago* has been applied in the steroid area. In a first report [26], this enzyme (simply prepared by grinding the acetone-dried mycelial powder in water and removing the mycelial debris by centrifugation) was employed to chlorinate and brominate steroidal β-diketones and aβ-ketolactone. The substrates used were 16-ketoprogesterone (**28**), 16-keto-A-norprogesterone (**29**), and 15-keto-1-dehydrotestololactone (**30**). After a short reaction time (1 h), compounds **28a** (or **28b**), **29a** (or **29b**), and **30a** were obtained in about 50% yield . Only starting material could be detected when the enzyme preparation was not added to the reaction mixture. Moreover, steroids without enolizable groups [e.g., progesterone (**31**)], were unaffected by the enzyme halogenating conditions.

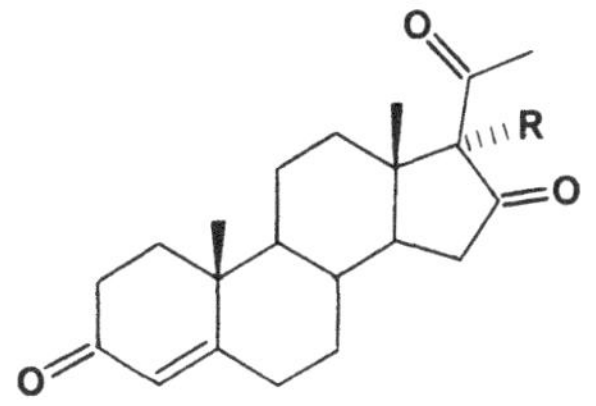

R = H	28
R = Br	28a
R = Cl	28b

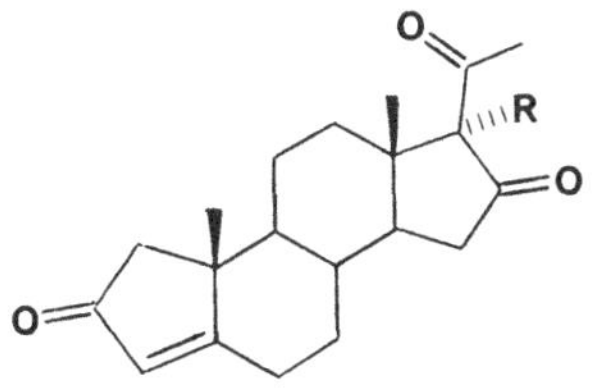

R = H	29
R = Br	29a
R = Cl	29b

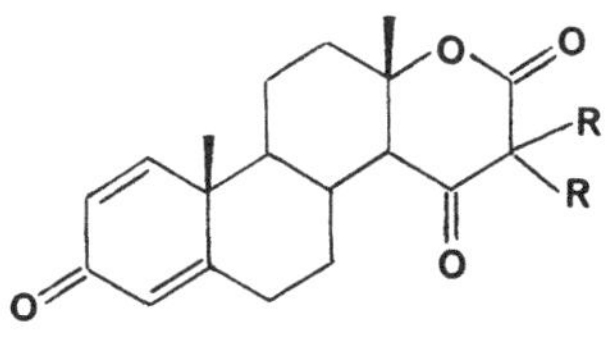

R = H	30
R = Br	30a

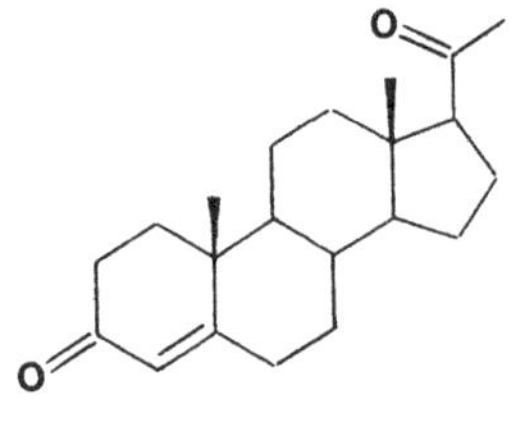

31

In a second report [27], another type of enolizable β-dicarbonyl system was used as a substrate for halogenation. For example, when 2-hydroxy-methylene-17β-hydroxy-5α-androstan-3-one (**32**) was incubated for 30 minutes at room temperature in a buffer solution (containing 5% DMSO to dissolve the steroid, the enzyme, hydrogen peroxide, and potassium bromide), the product 2α-bromo-17β-hydroxy-5α-androstan-3-one (**33**) was isolated in 20-25% yield.

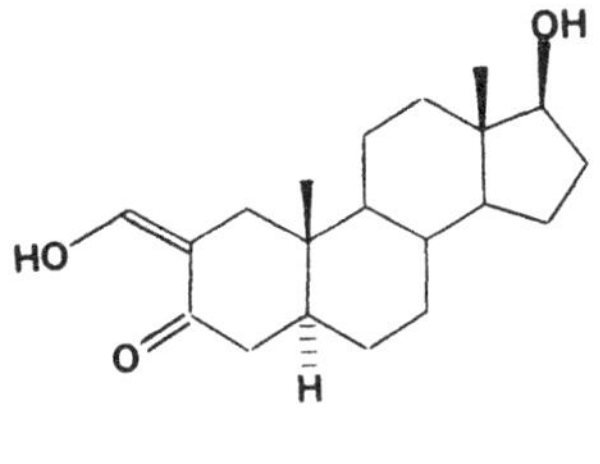

32

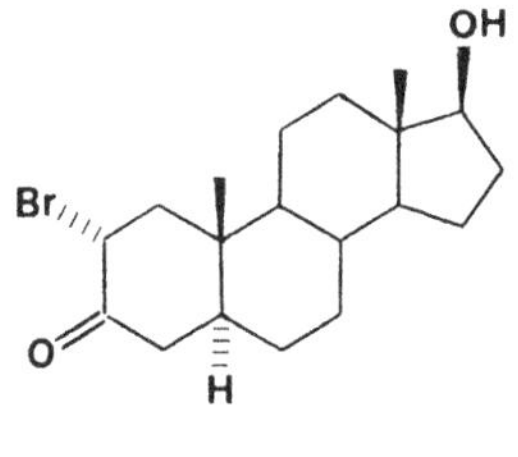

33

A third communication [28] reported enzymatic reactions with the isolated double bonds of 9(11)-dehydroprogesterone (**34**), pregnenolone (**35**), and pregnenolone acetate (**36**). In this case, haloperoxydase catalyzed the formation of the bromohydrin **37** (48% yield) or of the epoxides **38** (22% yield) and **39** (8% yield).

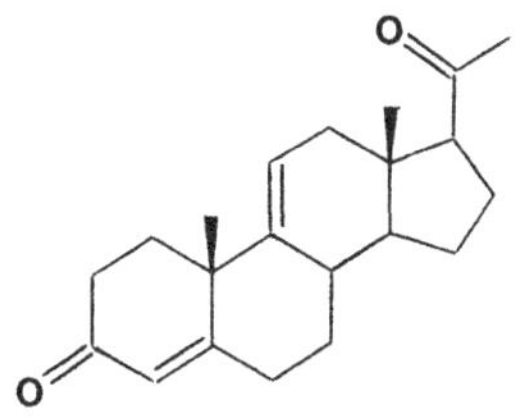

34

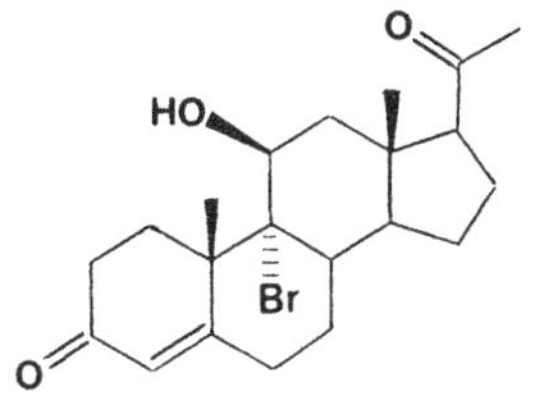

37

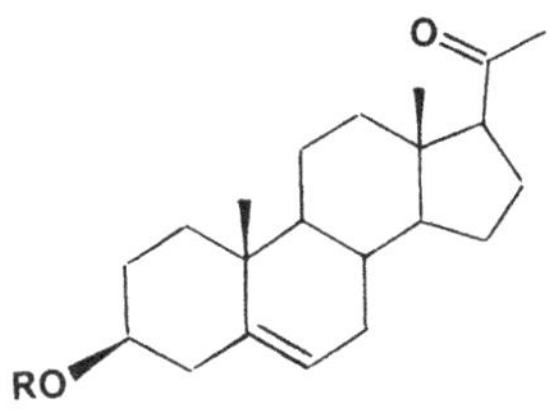

R= H 35
R= Ac 36

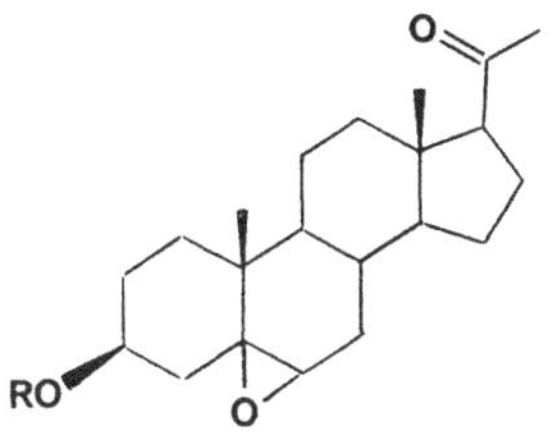

R=H 38
R=Ac 39

Monooxygenase

Some bacteria, capable of growth on aliphatic molecules, contain enzymes that can catalyze Baeyer–Villiger oxidations. These enzymes, known as monooxygenases, are involved in the catabolic pathways, providing simpler carbon units from the breakdown of acyclic and alicyclic ketones [29].

Prairie and Talalay [30] succeeded in obtaining an enzyme preparation from *Penicillium lilacinum* capable of converting progesterone (31) into testolactone (**41**) via androst-4-ene-3, 17-dione (**40**) (Fig. 5.6). Two steroid-in-

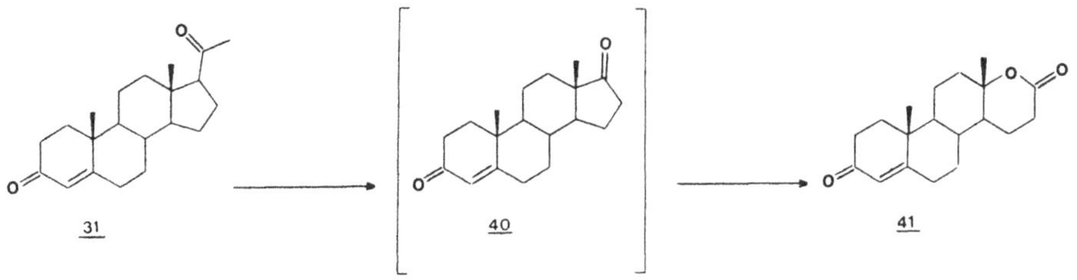

Figure 6 Conversion of progesterone (**31**) to testololactone (**41**) catalyzed by partially purified enzyme systems derived from *Penicillium lilacinum* [30].

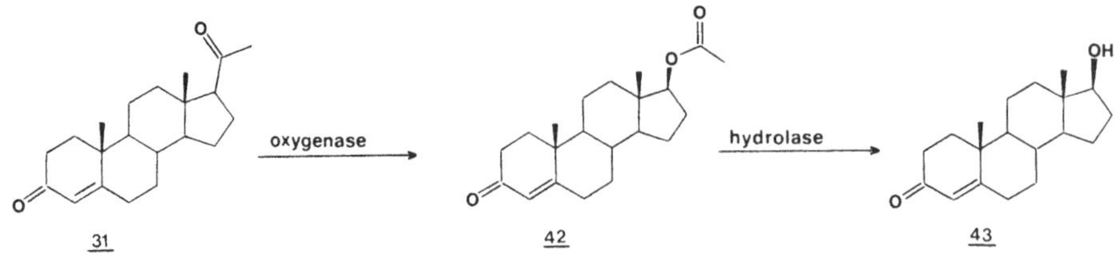

Figure 7 Conversion of testosterone (**31**) to testolocatone (**41**) catalyzed by partially purified enzymatic preparations from *Cylindrocarpon radicicola* [31].

duced enzymes were involved in this transformation, a 17β-HSDH and a lactone forming enzyme, which showed an absolute requirement for NADPH and molecular oxygen.

The same transformation was investigated by Rahim and Sih [31]. They studied the mechanism of pregnane side chain cleavage with the use of partially purified preparations of *Cylindrocarpon radicicola* and isolated a steroid-inducible oxygenase, which was capable of inserting an oxygen atom between C-17 and C-20 of progesterone (**31**), resulting in the formation of testosterone acetate (**42**). In the intact cells or in a crude cell extract, an esterase subsequently hydrolyzed **42** into testosterone (**43**) and acetic acid (Fig. 5.7). This oxygenase also showed an absolute requirement for NADPH and molecular oxygen. Its substrate specificity was also investigated, and the results are reported in Table 5.4.

Table 5.4 Substrate Specificity of Purified Progesterone Oxygenase [31]

Steroid	Relative activity
17α-Hydroxy-3,20-dioxo-4-pregnene	100
3,20-Dioxo-4-pregnene (progesterone)	47
21-Hydroxy-3,20-dioxo-4-pregnene	30
$11\beta,17\alpha,21$-Trihydroxy-3,20-dioxo-4-pregnene	0
17α-Methyl-3,20-dioxo-4-pregnene	2
20β-Hydroxy-3-oxo-4-pregnene	0
3,20-Dioxo-4,16-pregnadiene	49
$16\alpha,17\alpha$-Epoxy-3,20-dioxo-4-pregnene	84
$16\alpha,17\alpha$-Epoxy-20α-hydroxy-3,20-dioxo-4-pregnene	0

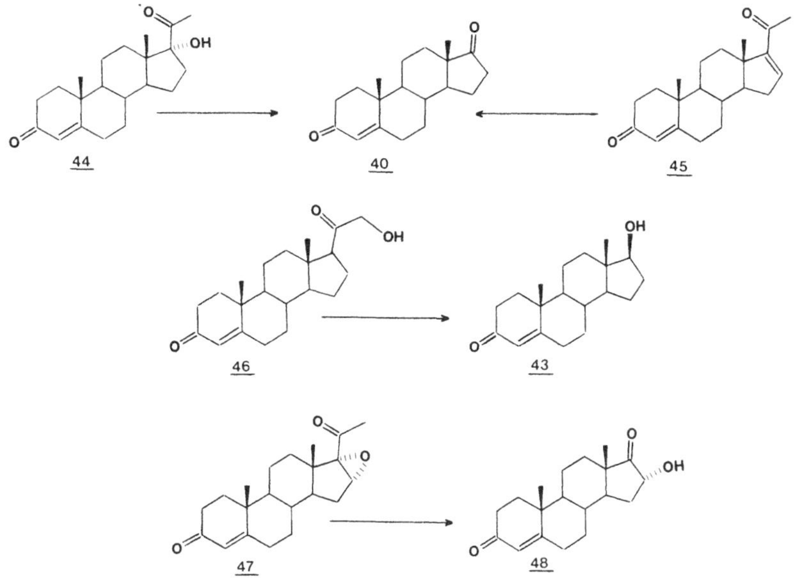

Figure 8 Conversions catalyzed by crude enzymatic preparations from *Cylindrocarpon radicicola* [31].

Some preparative-scale experiments were also carried out with 50 mg of different substrates, dissolved in DMF and then added to a buffer solution containing the active enzyme system. In this way 17α-hydroxypregn-4-ene-3,20-dione (**44**) was transformed into androst-4-ene-3,17-dione (**40**) and so was pregna-4,16-diene-3,20-dione (**45**); in addition, deoxycorticosterone (**46**) was converted into testosterone (**43**) and $16\alpha,17\alpha$-oxidopregn-4-ene-3,20-dione (**47**) into 16α-hydroxyandrost-4-ene-3,17-dione (**48**) (Fig. 5.8).

Glycosidase

Glycosidases, a class of specific hydrolases for sugar derivatives, have been widely used to liberate aglycons of diverse structures from their corresponding glycosides. Under adequate conditions, the reversed reactions can also be obtained. For instance, in 1936 Veible [32] prepared glucosides by an enzymatic method, using emulsin, glucose, and alcohol in water. Because this was an equilibrium reaction, it required a high concentration of alcohol to obtain the glucoside and consequently it was suitable only for the synthesis of lower alkyl glucosides. Quite recently, however, Satoh and coworkers [33,34] reported the synthesis of variously substituted and chemically unstable glycosides using a β-galactosidase from *Aspergillus oryzae* in 50% aqueous–acetonitrile solutions.

This enzyme catalyzed a transglycosilation between phenyl-β-galacto-pyranoside and several genins unstable toward acids and bases, such as gitoxigenin (**49**). The high volume percentage of organic cosolvents (which partially prevented the hydrolysis of the product glycosides), as well as low reaction temperatures and/or low reaction times, allowed the isolation of the corresponding 3-O- cardiac glycosides in 25–74%yields (Fig. 5.9).

ENZYMATIC REACTIONS IN BIPHASIC SYSTEMS

As we pointed out earlier, the progressive denaturation of enzymes is the major drawback associated with the addition of water-soluble cosolvents to the reaction mixtures. Furthermore, the use of cosolvents does not avoid the problem of enzyme inhibition, which is exerted, in many cases, by high concentrations of substrate or product.

However the situation becomes substantially different with the use of organic solvents, which are practically immiscible in water, or miscible only poorly. In this case a biphasic system consisting of water and an organic solvent is established [35-37]. In such a system, the water contains enzymes and hydrophilic cofactors, and the organic solvent contains the hydrophobic substrates. On shaking or stirring, transfer of substrates from the organic to the water phase takes place. The substrates undergo enzyme–catalyzed reactions, and products are formed and eventually return to the organic phase. In these systems, the concentration of solvents in water is low and is not dependent on the ratio of the two phases—even if the volume of the organic phase is much greater than that of the aqueous phase— and the inhibitory and denaturing effects are much weaker that those induced by comparable concentrations of miscible solvents. Enzymes that have been employed for the transformation of steroids mainly belong to the dehydrogenase group, but isomerase, oxidase, and laccase have also been used.

Hydroxysteroid Dehydrogenase

Regio- and stereospecific oxidoreductions of the hydroxyl–keto groups of steroids catalyzed by NAD(P)-dependent HSDH in two-phase systems have been described in several papers [38]. In fact the combined use of biphasic milieu and effective enzymatic methods for cofactor regeneration has made these re-

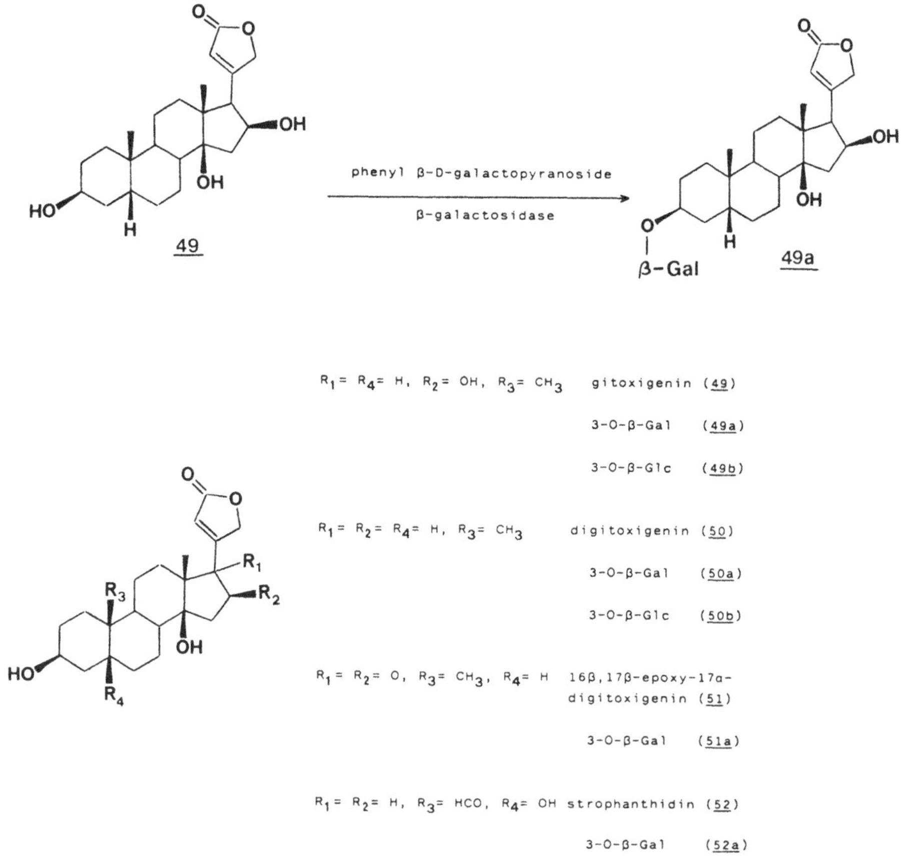
phenyl β-D-galactopyranoside
β-galactosidase
49
49a
β-Gal
$R_1 = R_4 = H, R_2 = OH, R_3 = CH_3$ gitoxigenin (49)
3-O-β-Gal (49a)
3-O-β-Glc (49b)
$R_1 = R_2 = R_4 = H, R_3 = CH_3$ digitoxigenin (50)
3-O-β-Gal (50a)
3-O-β-Glc (50b)
$R_1 = R_2 = O, R_3 = CH_3, R_4 = H$ 16β,17β-epoxy-17α-digitoxigenin (51)
3-O-β-Gal (51a)
$R_1 = R_2 = H, R_3 = HCO, R_4 = OH$ strophanthidin (52)
3-O-β-Gal (52a)

Figure 9 Enzymatic synthesis of cardiac glycosides catalyzed by β-galactosidase from *Aspergillus oryzae* [33].

actions particularly suitable for preparative -scale transformations. Some examples are depicted in Figure 5.10.

Now we turn to criteria that must be taken into consideration for selecting the most suitable organic solvent and the operational conditions.

Solubility and Partition of Reagents and Products Between Phases

The choice of the solvent should reflect consideration of several factors and it is often a compromise, since it is unlikely that one solvent will possess all the ideal characteristics (high solubilizing capacity, good partition coefficient with water, noninactivating action on the enzyme). This is illustrated in Table

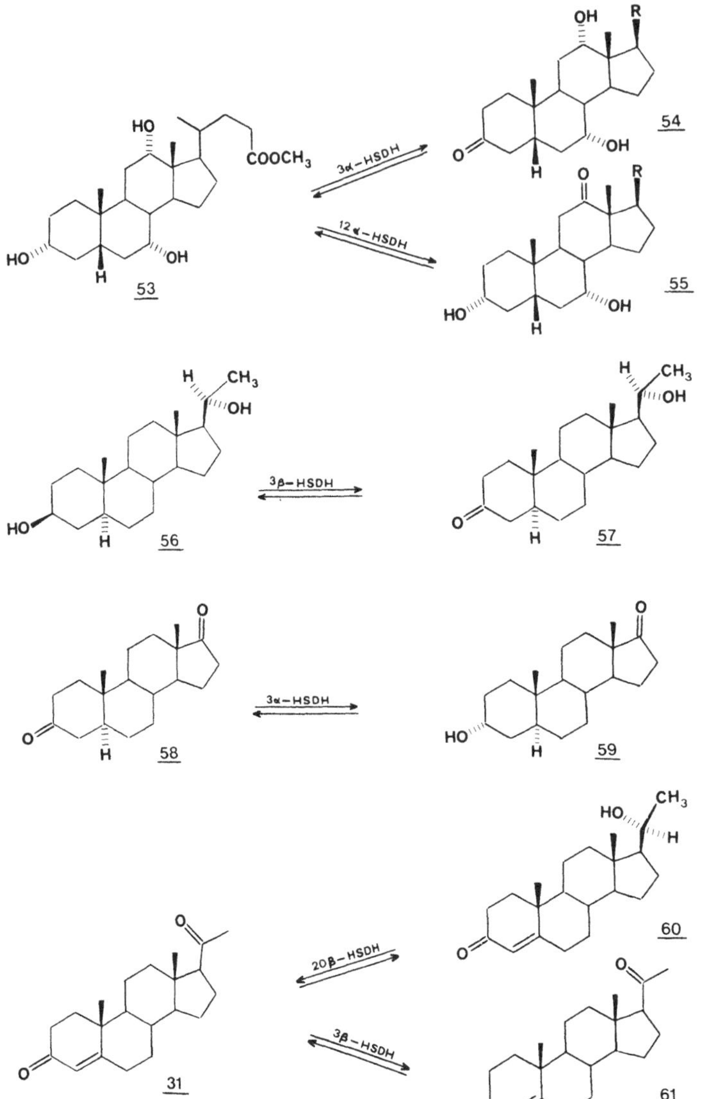

Figure 10   Examples of regiospecific oxidations and regio- and stereospecific reductions of apolar steroids catalyzed by hydroxysteroid dehydrogenases in biphasic systems [38].

Table 5.5 Effect of Organic Solvents on 20β-Hydroxysteroid Dehydrogenase Catalyzed Reduction of Cortisone (17α,21-dihydroxy-3,11,20-trioxo-4-pregnene) [39]

Organic solvent	Relative enzymatic activity	Relative enzymatic stability	Cortisone solubility (g/100mL)	Conversion (%)
Control (water)	100	100		
Hexane	97	95	0.002	< 5
Isooctane	100	91		< 5
Carbon tetrachloride	100	74	0.004	< 5
Chlorobenzene	95	81	0.030	15
Trichloroethylene	90	76		10
Diethyl ether	38	52	0.017	10
Butyl acetate	48	58	0.160	100
Ethyl acetate	29	48	0.270	90

5.5, which shows the yields of the 20β-HSDH-catalyzed reduction of cortisone in different solvents [39]. It can be seen that solvents that partially destabilize and inhibit the enzyme but solubilize large amounts of substrate are more productive than solvents that do not affect enzyme activity and stability but solubilize only small amounts of substrate. Ethyl acetate and butyl acetate were found to be the most suitable solvents in several steroid transformations.

Transfer Between Phases

The transfer of reagents between the aqueous and organic phases is easily achieved by shaking or stirring. When free enzymes are employed, very high rates of shaking must be avoided because enzymes can become destabilized [40]. In the process, the resistance to transfer of substrates or products from one phase to another should be negligible compared with the diffusional resistance in each of the two phases. The situation is more complex when immobilized enzymes are used, and the penetration of reagents into the matrix may be rate limiting. The effect of pH on enzymatic catalysts in two-phase systems is the same as that in aqueous buffers. The reaction rate is dependent on the concentration of the substrate in the water phase rather than on its total concentration. This indicates that catalysis occurs in the water phase and rules out the hypothesis that enzymatic transformations occur at the interfacial area.

Table 5.6 Effect of Ethyl Acetate on the Kinetic Parameters of β-HSDH, 20β-HSDH, 12α-HSDH, and 3α-HSDH [41]

Enzyme	Solvent concentration (M)	Substrate	K_M (μM)	V_{max}, Relative
β-HSDH	0	Estradiol	2.1	100
	0.6	Estradiol	9.2	100
20β-HSDH	0	Cortisone	29.0	100
	0.7	Cortisone	95.0	95
12α-HSDH	0	Cholic acid, methyl ester	59.2	100
	0.7	Cholic acid, methyl ester	390.0	100
	0	Cholic acid	64.0	100
	0.7	Cholic acid	420.0	125
3α-HSDH	0	Androsterone	2.8	100
	0.7	Androsterone	3.4	21
	0	Cholic acid, methyl ester	3.0	100
	0.7	Cholic acid, methyl ester	6.8	37
	0	Cholic acid	4.3	100
	0.7	Cholic acid	31.0	60

Enzyme Kinetics

The effect of various water-immiscible organic solvents on the activity of HSDH has been examined. In particular, the effects of ethyl and butyl acetate on the K_M and V_{max} values of the enzymes have been studied [41]. Table 5.6 shows that ethyl acetate behaves as a competitive inhibitor (increases the K_M) for steroid substrates in the cases of β-HSDH, 20β-HSDH, and 12α-HSDH, whereas with 3α-HSDH there is a mixed competitive and noncompetitive (decrease of V_{max}) inhibition. The decrease of V_{max} values is always a serious drawback from a practical point of view, since even at high enzyme concentrations of the substrate, the enzyme is not fully active. Ethyl acetate did not substantially influence the K_M values of the HSDH for NAD(P)(H) and also did

Table 5.7 20-β-Hydroxy-3-oxo-4-pregnene as Product Inhibitor of 20β-HSDH
Using Progesterone as Substrate [41]

Inhibitor concentration (μM)	Buffer			Buffer + 0.7 M AcOEt		
	K_M (μM)	$V_{max,}$ relative	K_i (μM)	K_M (μM)	$V_{max,}$ relative[a]	K_i (μM)
0	3.5	100		9.5	100	
42	3.5	45	35	9.7	63	72
84	3.6	30	33	9.3	46	71

[a]In 0.7 M ethyl acetate, V_{max} is 140% of that in buffer.

not affect the activity of the enzymes used for the regeneration of the nicotinamide cofactors.

Noncompetitive inhibition by steroid products is reduced by the organic solvent not only because it reduces the concentration of the product in the aqueous phase, but also because it increase its K_i value; that is, it reduces the affinity of the product for the enzyme [41]. Table 5.7 shows the inhibition pattern of 20β-HSDH in the presence or absence of ethyl acetate. The K_i value of the product is doubled in the presence of the organic solvent, which means that in the presence of the organic solvent, the concentration of product that will lower to 50% the apparent concentration of the enzyme will be two times higher than in buffer alone.

Reactor Evaluation

Carrea and coworkers compared the performances of various reactors: free enzymes in shaken vessels, covalently immobilized enzymes in shaken vessels, covalently immobilized enzymes in fixed-bed reactors, enzymes adsorbed in fixed-bed reactors [41]. The fastest reaction rate was obtained with free enzymes in a shaken vessel, and the slowest with immobilized enzymes in a shaken vessel. Fixed-bed reactors employing covalently immobilized enzymes were the most advantageous for repeated transformations, because of increased enzyme stability, high coenzyme turnover number, and satisfactory mass transfer rates. Free enzymes, which are easier to handle, were preferable for occasional use and small-scale preparations.

Isomerase

The isomerization of 5-androsten-3,17-dione (**62** to 4-androsten-3,17-dione (**40**) catalyzed by $\Delta^5 \rightarrow \Delta^4$-3-oxosteroid isomerase has been investigated under biphasic conditions [42]. The dependence of the reaction rate on organic solvent concentration , substrate concentration, and the kinetic profile at various pH values was considered. The results showed that for ethyl acetate, butyl acetate, and diethyl ether the apparent K_M value increased with organic solvent concentration, while the V_{max} value remained constant, suggesting that some kind of competitive inhibition of the organic solvent toward the substrate was taking place.

Laccase

Laccases (*p*-diphenol: O_2 oxidoreductase) are copper-containing enzymes that catalyze the oxidation by molecular oxygen of a variety of substrates. A fungal laccase (from *Polyporus versicolor*) catalyzed the oxidation by molecular oxygen of steroid hormones with a phenolic group in the A ring [43]. The oxidation was performed in biphasic systems. With several organic solvents the laccase displayed full activity over a long period of time (up to several days), and the best results were obtained in ethyl acetate, butyl acetate, diethyl ether, and methyl ethyl ketone. Comparison of the data obtained on a variety of compounds indicated the need for a phenolic group on the steroid for the reaction. Using estradiol (**63**) as substrate at least six oxidation products were formed. Two of them were present in larger amounts and, upon isolation and spectroscopic characterization, were found to be dimeric compounds derived from the oxidative condensation of two phenolic molecules.

ENZYMATIC REACTION IN REVERSED MICELLES

Reversed micelles are tiny droplets of water stabilized by surfactants in a bulk, water-immiscible organic solvent. As has been shown by several authors [44], reversed micelles produce a microenvironment that protects enzymes against denaturation by the surrounding organic solvent and enables them to convert hydrophobic compounds present in the bulk organic phase. These reversed micelles possess an enormous interfacial area (100 m²/mL), and the distance between the interface and the enzyme in the water pools is as small as possible. Thus the effect of diffusional limitation is greatly reduced.

The use of dehydrogenases for reducing water-insoluble steroids in such milieus was studied by Laane and coworkers [45], who investigated the regio- and stereospecific reduction of 20-ketosteroids to their corresponding 20β-hydroxy derivatives. Progesterone (**31**) was initially chosen as a model substrate for 20β-HSDH. A multienzyme system, consisting of hydrogenase, lipoamide dehydrogenase, and 20β-HSDH, was enclosed in reversed micelles. In such a system H_2 was used by hydrogenase to reduce MV^{2+} to MV^+. Two of these radicals were consumed by lipoamide dehydrogenase to regenerate NADH from the NAD formed by 20β-HSDH during progesterone reduction (Fig.5.11). The coupling of these reactions led to the recycling of both MV^{2+} and NAD while H_2 was consumed (Fig.5.12). As expected, the penetration of the steroid into the micelles was influenced strongly by the composition of the system and by the nature of the cosurfactant. When all the steroid had been transformed, the enzymes and other water-soluble components

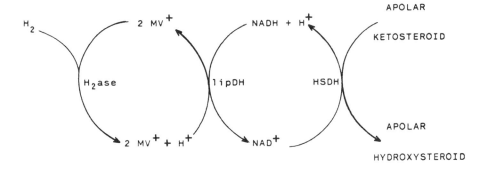

Figure 11 H_2-driven regeneration of NADH and subsequent reduction of an apolar steroid in reversed micelles [45] H₂ase, hydrogenase; lipDH, lipoamide dehydrogenase; HSDH, hydroxysteroid dehydrogenase; MV, methyl viologen.

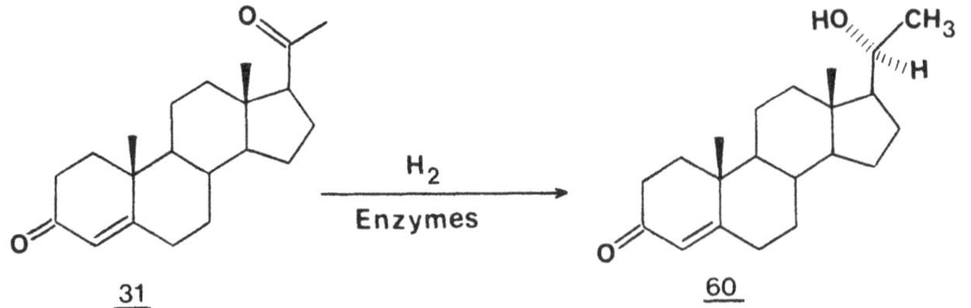

Figure 12 Net reaction for the reduction of progesterone (**31**) by the multienzyme system shown in Figure 5.11 [45].

could be recycled in fairly good yield (80%), and the product was isolated from the organic phase after precipitation of the surfactant with acetonitrile.

A more detailed investigation of the parameters involved in this kind of steroid transformation was published later [46]. It has been demonstrated that the solubility of the substrate can be optimized by tuning the polarity of the bulk organic phase and the polarity of the interphase to the polarity of the steroid (the logarithm of the standard partition coefficient, log P, was used as a quantitative measure of polarity). Nevertheless, there is a limit to the amount of substrate that can be solubilized in reversed micelles. For steroids such as progesterone, this limit is approximately 0.1–0.2 M.

ENZYMATIC REACTION IN ORGANIC SOLVENTS

In all the reactions thus far presented, the enzyme molecules were located in essentially aqueous milieus. It was not surprising, therefore, that the biocatalysts remained active. However quite recently it has been observed that enzymes can work even in anhydrous organic media (where the term "anhydrous" means that the water content is below 0.1% v/v, which is the sensitivity limit of the classic Karl Fischer method). In these cases, the solid enzymes are simply stirred into an organic solvent and used as a catalyst in the form of an undissolved suspension.

The behavior and the new properties of enzymes in organic solvents have been elucidated in depth by Klibanov and coworkers [47, 48]. Among the different interesting characteristics, it is of practical importance that enzymes can catalyze reactions that are virtually impossible in water. When placed in organic solvents for example, lipases and proteases can catalyze transesterifica-

Table 5.8 Reactivities of Various Hydroxysteroids in the Acylation Catalyzed by *Chromobacterium viscosum* Lipase and Subtilisin in Anhydrous Acetone [50]

Steroid	Structure	Initial Rate (μmol/h)	
		Lipase	Sutilisin
$3\beta,17\beta$-Dihydroxy-5α-androstane	**64**	3.30	0.63
$3\alpha,17\beta$-Dihydroxy-5α-androstane	**65**	0	0.53
$3\beta,17\beta$-Dihydroxy-5β-androstane	**67**	0	0.41
$3\alpha,17\beta$-Dihydroxy-5β-androstane	**66**	0	0.63
3β-17β-Dihydroxy-5-androstene	**68**	1.72	0.67
3β-17β-Dihydroxy-4-androstene	**69**	0	0.32
17β-Estradiol	**63**	0	0.63
3β-Hydroxy-5α-pregnane	**70**	6.26	0.06
3β-Hydroxy-5α-cholanic acid methyl ester	**71**	8.32	0
3β-Hydroxy-5-cholenic acid methyl ester	**72**	2.75	0
3β-Hydroxy-5-cholestene (cholesterol)	**73**	3.00	0
$3\beta,20\beta$-Dihydroxy-5α-pregnane	**74**	4.52	0.10
$3\beta,20\alpha$-Dihydroxy-5α-pregnane	**56**	4.34	0.55

tion, esterification, aminolysis, acyl exchange, and thiotransesterification [49]. In water all these reactions are suppressed by the usual hydrolytic activity, and therefore they do not occur to any appreciable extent.

These hydrolytic enzymes have been employed for the regioselective esterification of hydroxysteroids. In a first report [50], two commercially available hydrolases (out of eight tested), *Chromobacterium viscosum* (*Ch. v*) lipase and *Bacillus subtilis* protease (subtilisin), have been found to esterify the model dihydroxysteroid 5α-androstane-$3\beta,17\beta$-diol (**64**) in dry acetone. These enzymes acrlated the two hydroxyl groups in **64** with opposite regioselectivities: while *Ch. v*. lipase reacted exclusively with OH in the C-3 position, subtilisin displayed a marked preference for the C-17 hydroxyl (the chemical reactivities of these two hydroxyl groups in **64** were comparable). The reactivity of various other hydroxysteroids (**64–74**) with *Ch. v*. lipase and subtilisin was examined and the results, reported in Table 5.8, afforded several conclusions.

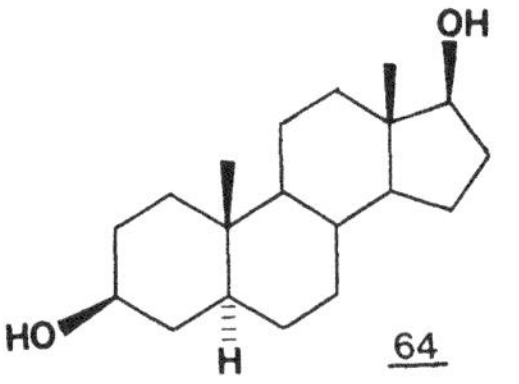

64

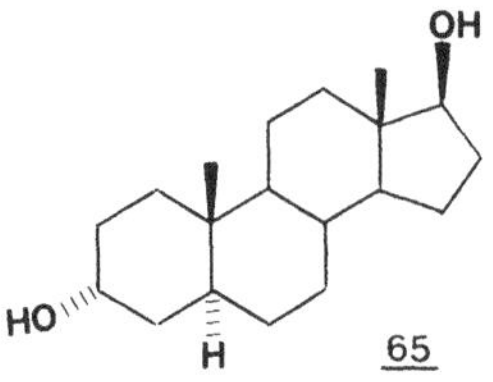

65

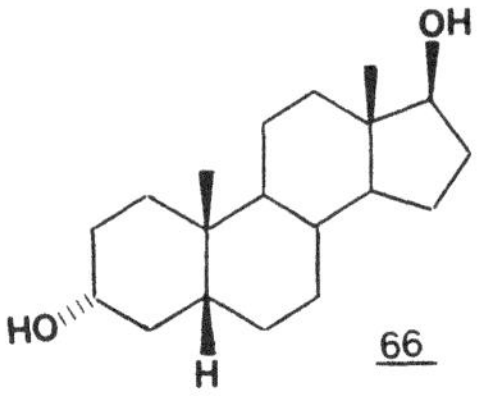

66

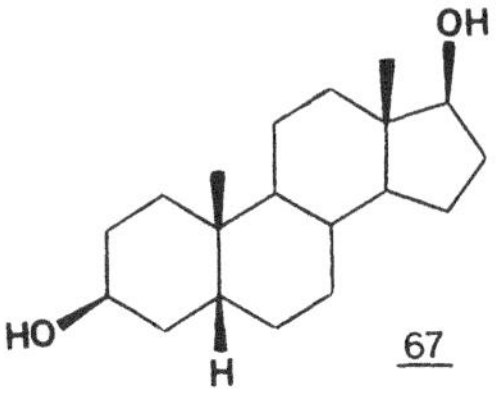

67

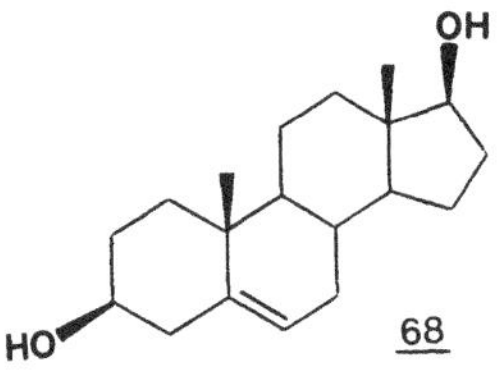

68

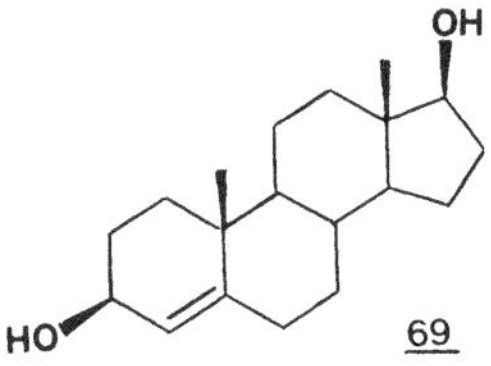

69

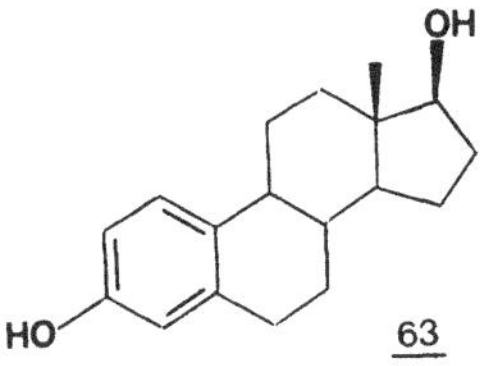

63

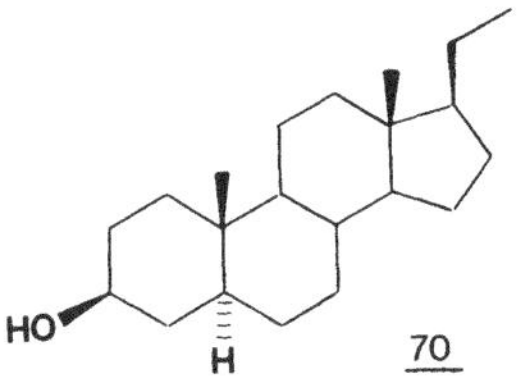

70

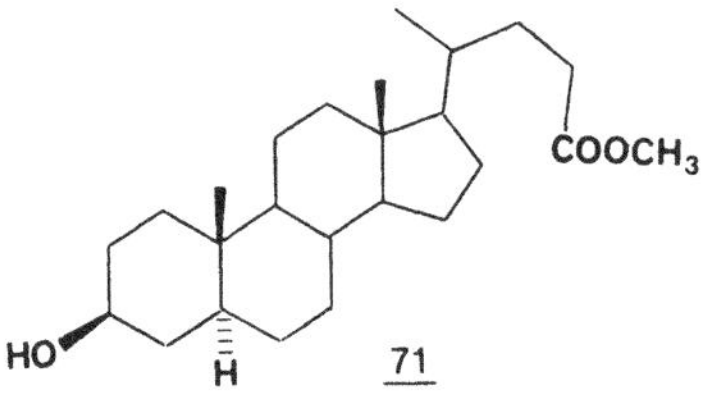

71

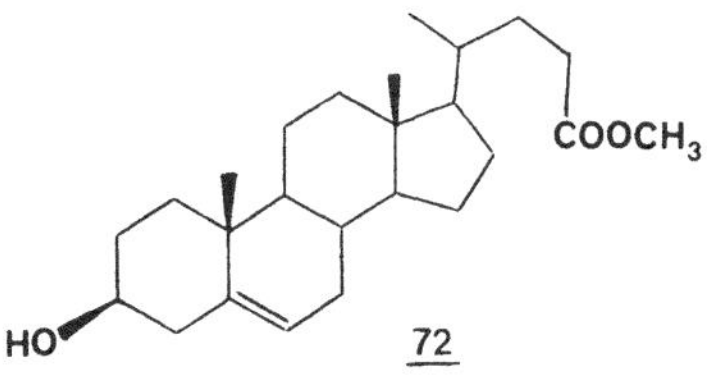

72

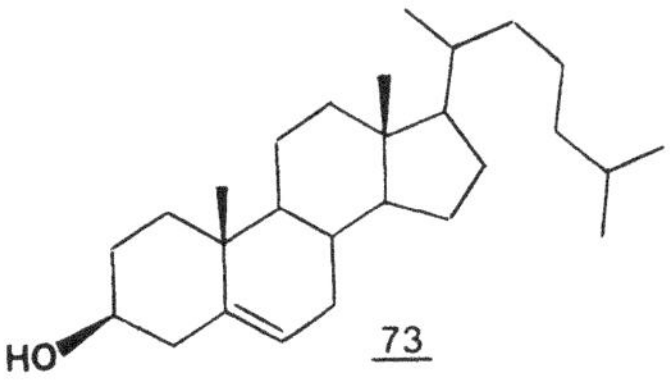

73

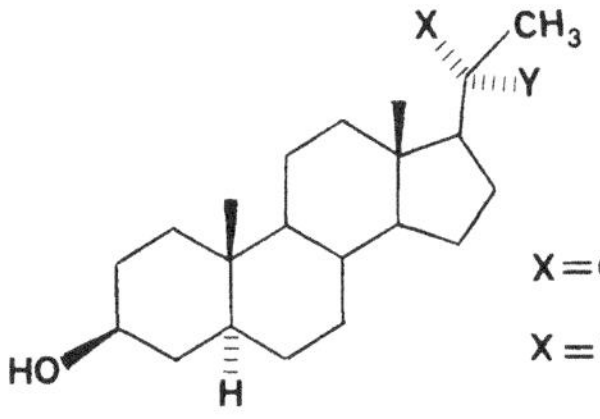

X=OH, Y=H   74

X=H. Y=OH   56

Comparison of the reactivities of **64-67** toward *Ch. v.* lipase indicated that the substrate must have the A/B ring fusion in the trans configuration and the C-3 hydroxyl group in the equatorial (β) position. The sensitivity of the enzyme to the immediate environment of the steroids C-3 OH group was supported by the finding that the introduction of a double bond in the B ring was tolerated, while in the A ring it was not (compounds **68** and **69**, respectively). The phenolic group in **63** was unreactive. The compounds with the altered side chain (**70-74**) were very reactive with lipase as long as the C-3 hydroxyl and C-5 hydrogen remained in β and α configurations, respectively.

Conversely, in the case of subtilisin, changes in the A or B ring did not dramatically affect the reactivity of the steroid (compounds **63-69**). Very low reaction rates of compounds **70-73** were consistent with the earlier conclusion that subtilisin strongly prefers the C-17 or side chain hydroxyl group over that in the C-3 position.

Therefore, both *Ch. v.* lipase and subtilisin required the substrate's hydroxyl group and its immediate surroundings to be in a certain orientation, while they were virtually indifferent to the rest of the steroid structure.

Enzymatic esterifications of some of the steroids in Table 5.8 were scaled up, and the data obtained are presented in Table 5.9 *Ch. v.* lipase catalyzed acylation of steroids afforded pure 3β-monoester products with 83-85% isolated yields. In the case of subtilisin, the process was considerably slower, but it was still of preparative value, and pure 17β-monoesters were produced with 60% isolated yields.

The same strict regioselectivity prevented the two enzymes from acylating hydroxysteroids with different stereochemical characteristics--- for instance,

Table 5.9 Preparative Regioselective Acylation of Dihydroxysteroid Catalyzed by *Chromobacterium viscosum* Lipase and Subtilisin in Anhydrous Acetone [50]

Steroid	Enzyme	Conversion (%)	Isolated yield (% theory)	Product
64	Lipase	96	83	3β-O-acyl-**64**
64	Subtilisin	79	60	17β-O-acyl-**64**
68	Lipase	93	84	3β-O-acyl-**68**
68	Subtilisin	85	63	17β-O-acyl-**68**
74	Lipase	98	85	3β-O-acyl-**74**

Table 5.10 Reactivities of Various Hydroxysteroids in the Acylation Catalyzed by *Candida cyllindracea* Lipase in Anhydrous Benzene [51]

Steroid	Structure	Initial rate (μmol/h)
3β-Hydroxy-17-oxo-5α-androstane		4.74
3β-Hydroxy-17-oxo-5β-androstane		0.13
3α-Hydroxy-17-oxo-5β-androstane		0.57
3α-Hydroxy-17-oxo-5α-androstane		0
3β-Hydroxy-17-oxo-5-androstene		4.15
3β-Hydroxy-20-oxo-5-pregnene		8.08
17β-Hydroxy-3-oxo-5α-androstane		0.75
17β-Hydroxy-3-oxo-5β-androstane		0
20β-Hydroxy-3-oxo-4-pregnene		0
Deoxycholic acid, methyl ester	**75**	4.59
Lithocholic acid, methyl ester	**71**	1.58
Chenodeoxycholic acid, methyl ester	**76**	0.94
Ursodeoxycholic acid, methyl ester	**77**	0.67
Cholic acid, methyl ester	**53**	2.47
$7\alpha,12\alpha$-Dihydroxy-3-oxo-5β-cholanoic acid, methyl ester	**78**	0

deoxycholic acid methyl ester (**75**). However, simply by changing the solvent and moving from acetone to the less hydrophilic benzene, a different lipase (from *Candida cylindracea, Cn. c*) was able to regioselectively acylate several bile acid derivatives. [51].

The substrate specificity of *Cn. c* lipase toward different steroid skeletons was also investigated. The results, depicted in Table 5.10, showed that the stereochemcal requirements of this enzyme are less strict than those of *Ch. v.* lipase. Bile acids methyl esters **53, 72, 76-77** were also accepted by the enzyme's active site, while no conversion was observed with the 3-keto derivative **78**. Reactions were scaled up with compounds **53, 71, 75-77**, and the corresponding 3α-O-acyl derivatives were isolated in 70-85% yield.

A third paper [52] reported the use of enzymatic transesterification as a mild methodology for the regioselective deprotection of polyacylated steroids. *Candida cylindracea* lipase catalyzed transesterification of steroid esters with

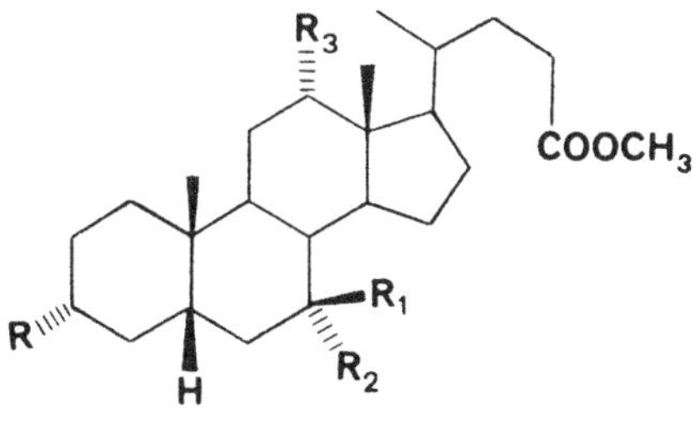

COMPOUND	R	R	R$_2$	R$_3$
75	OH	H	H	OH
71	OH	H	H	H
76	OH	H	OH	H
77	OH	OH	H	H
53	OH	H	OH	OH

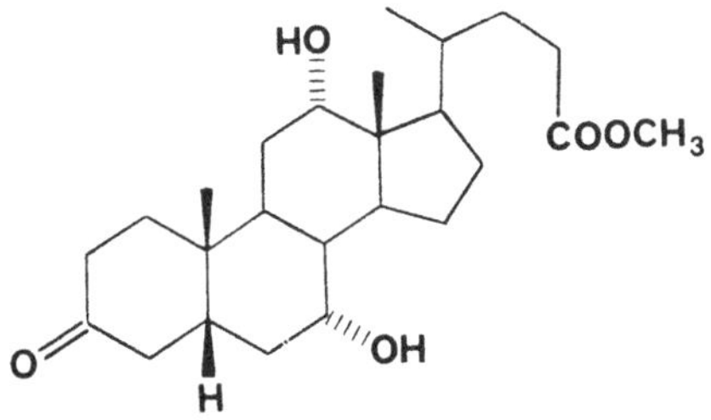

78

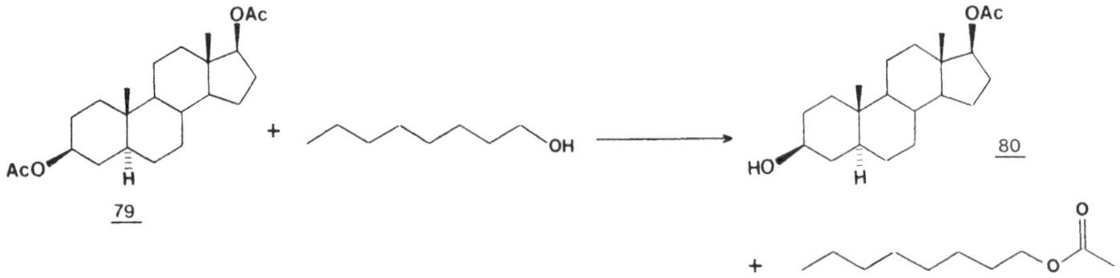

Figure 13 Regioselective deprotection of 3β,17β-diacetoxy-5α-androstane (**79**) catalyzed by *Candida cylindracea* lipase in dry isopropyl ether [52].

octanol in either isopropyl ether or acetonitrile. For example, Figure 5.13 reports the conversion of 3β,17β-diacetoxy-5α-androstane **79** to 17β-acetoxy-5α-androstan-3β-ol **80** (68% yield).

Several different steroid esters were used as substrates; among them, the 3α-acetoxy, 17β-acetoxy, and 19-acetoxy resisted the saponification, whichever the A/B ring junction was.

ENZYMATIC REACTIONS IN SUPERCRITICAL CARBON DIOXIDE

Supercritical fluids are appealing nonaqueous solvents for enzymatic catalysis, mainly because of their low viscosities and the high diffusivities of the substrates dissolved in these peculiar media. Moreover, large solubility changes may result from small changes in temperature or pressure near the critical point, thus facilitating the downstream separation steps.

Among the different gases, carbon dioxide—which is inert, nonflammable, inexpensive, nontoxic, and has a critical temperature of 31.1°C—is particularly attractive for biochemical processes. Up to now, enzymatic activity in supercritical CO_2 has been demonstrated with alkaline phosphatase, polyphenol oxidase, lipases from different sources, and the protease subtilisin.

Modification of hydrophobic steroids may also be of particular interest and, as a first example, Blanch and coworkers examined the enzyme-catalyzed kinetics of cholesterol (**73**) oxidation by molecular oxygen in supercritical CO_2 [53,54]. The enzyme employed, cholesterol oxidase from *Glecocysticum chrysocreas*, catalyzes the oxidation of **73** to 4-cholesten-3-one (**81**), a precursor of interest in the pharmaceutical production of androst-1,4 -diene-3,17-dione (quite interestingly, this enzyme has been employed in most of the solvent systems described up to now [55], thus making possible a comparison between them).

Cholesterol oxidase was covalently immobilized with glutaraldehyde to 50 μm nonporous glass beads and placed in a continuous flow, packed-bed reactor. A saturated stream of cholesterol in 9:1 mixture of CO_2 and O_2 was pumped through the reactor at a temperature of 35°C and at a pressure of 100 bar. Production of **81** was monitored by on-line UV spectroscopy at 242 nm. Under these conditions, the activity of cholesterol oxidase was retained for at least 3 days.

Quite interesting was the effect of cosolvents (MeOH, EtOH, acetone, *n*-butanol, isobutanol, *tert*-butanol), added to the reaction mixture with the initial aim of increasing cholesterol solubility. It was found that an increased solubility(obtained, e.g, with a small amount of methanol) did not necessarily lead to higher reaction rates. Conversely, substantial increases of enzyme activity were observed with cosolvents that were able to alter the structure of cholesterol aggregates in the reaction milieu (demonstrates by electron paramagnetic resonance spectra) [54]. These results, unpredictable a priori, point out the need for more detailed investigations of the properties of this unusual solvent and of its interaction with suspended enzymes, as our knowledge of this subject is still in its infancy.

ENZYMATIC ANALYSIS OF STEROIDS

Enzymes' keen substrate specificity makes those biocatalysts particularly suitable for analytical applications. Quantitative enzymatic determinations of numerous compounds of clinical, biochemical, and chemical interest have been reported in thousands of papers and patents. Most of these analytical protocols have been developed for water solutions, but some examples of enzymatic sensors employed in organic solvents or in gas phase systems have been described.

Several methods have been also optimized for the determination of specific steroid derivatives. A comprehensive overview has been published quite recently [56], reporting several detailed experimental examples with most of the enzymes cited earlier: hydrolases, hydroxysteroid dehydrogenases, oxidases, etc. Hydrolases (β-glucuronidase, sulfatase, cholesterol esterase) have been employed for the complete and specific hydrolysis of steroid conjugates, that is:

$$\text{steroid - glucuronide} + H_2O \longrightarrow \text{steroid alcohol + glucuronic acid}$$
$$(2)$$

The neutral steroids thus obtained were subsequently extracted with organic solvents and determined with the appropriate analytical protocol.

Specific hydroxysteroid dehydrogenases allowed the determination of neutral steroids and bile acid derivatives in serum or other physiological liquid. Taking bile acids as an example, all the methods exploited the regiospecific oxidation of the 3α-hydroxyl by 3α-HSDH [57]:

$$3\alpha \text{ - hydroxyl bile acid} + NAD \rightleftarrows 3 \text{ - oxo bile acid } + NADH + H$$
$$(3)$$

The reaction depicted in Equation (3) is reversible. However, at alkaline pH and with ketone-trapping agents such as hydrazine hydrate, the oxidation of 3α-hydroxy bile acids was practically quantitative. The exact amount of NADH formed could be determined by UV spectroscopy, fluorescence, colorimetry, or luminsecence.

In the spectrophotometric assay, NADH was measured by its absorption of light at 340 nm. Since the extinction coefficient of the reduced cofactor is known, the amount of steroid can be calculated without the use of an internal standard. This assay is reliable and convenient for relative pure samples, with a sensitivity of about 10^{-8} mol ($0.5\mu g$ of steroid).

Fluorimetric assay could be performed either by exploiting the native fluorescence of NADH at alkaline pH or, more efficiently, by coupling a diaphorase-catalyzed reaction:

$$NADH + H + \text{resazurin} \longrightarrow NAD + H_2 + \text{resorufin} \qquad (4)$$

Fluorescence of resorufin was excited at 565 nm and measured at 580 nm. This has a higher sensitivity (up to 10^{-11} mol) and is more suitable for determination of bile acids in serum [58].

Quite similar was the elaboration required with the third analytical method. In this case, however, diaphorase catalyzed the formation of formazan, a blue compound with an absorption maximum at 540 nm [59]:

$$NADH + H + \text{nitroblue tetrazolium salt} \longrightarrow NAD + \text{formazan} \tag{5}$$

The elimination of the need for preincubation of the sample and the possibility of using a simple spectrophotometer, are the advantages of this method compared with fluorimetry.

Finally, the bioluminescent assay determination of NADH was based on an enzymatic system consisting of an FMN-oxidoreductase and a luciferase that emits light in the presence of FMNH2 [60]:

$$NADH + H^+ + FMN \longrightarrow NAD + FMNH_2 \tag{6}$$

$$FMNH_2 + O_2 + RCHO \longrightarrow FMN + RCOOH + H_2O + h\nu \tag{7}$$

The sensitivity of the NADH assay was very high and, with enzymes coimmobilized on nylon tubes, as little as a picomole of standard was detected [61].

Cholesterol oxidase has been used to determine serum total cholesterol [56]. The most widely employed analytical protocol is described in Equations (8)-(10).

$$\text{cholesterol ester} + H_2O \longrightarrow \text{cholesterol} + \text{fatty acid} \tag{8}$$

$$\text{cholesterol} + O_2 \longrightarrow \Delta^4 - \text{cholestenone} + H_2O_2 \tag{9}$$

$$2H_2O_2 + \text{phenol} + 4 - \text{aminoantipyrine} \longrightarrow$$
$$4 - \text{benzoquinone} - \text{monoiminophenazone} + 4H_2O \tag{10}$$

The quinone imine dye showed a broad maximum centered at 500 nm, and the increase of its adsorbance was proportional to the quantity of total cholesterol.

The same scheme (Eq. 8-10) was found to be suitable for use in organic solvents, allowing accurate and reproducible determination of cholesterol in toluene [62]. In this case the two enzymes (cholesterol oxidase and peroxidase) had been adsorbed previously on glass powder, and p-ansidine was used as a chromogenic substrate for the peroxidase.

This approach is of potential interest for the determination of hydrophobic substrates because these substances can be preextracted from serum solutions with the appropriate organic solvent, thus greatly enhancing the concentration of the analyte above the sensitivity level.

CONCLUSIONS

To the best of the author's knowledge, this chapter has presented an overview of the literature published on enzymatic modification of steroid. It is surely plain, however, that hydroxylation reactions have been completely neglected. The reason is that, while the complex enzymatic mechanisms involved in these outstanding regio- and stereospecific transformations have been quite well elucidated, the extreme instability of the purified proteins has seriously inhibited their practical use. At the present, the results obtained with isolated enzymes are far from being competitive with the usual mode of fermentation. Moreover, side-directed mutagenesis has disclosed new horizons in this field. Both academic and industrial research groups are in fact actively involved in programs aimed at cloning the genes of the specific hydroxylases into new and more suitable microorganisms.

These new potentialities of the fermemtative approach suggest a final, and in a way obvious, consideration. The isolation and characterization of the enzymes devoted to specific biotransformations of the steroid molecule is always valuable from a scientific point of view. Moreover, new and a priori unpredictable results can be obtained by using "unnatural" reaction milieu, as has been pointed out in the preceding pages. However, just as it is always a matter of discussion whether biological systems are more suitable than the usual chemical procedures for a specific reaction, in the same way the choice of isolated enzymes (more or less purified) instead of whole cells must be considered case by case.

ACKNOWLEDGMENTS

I express my gratitude to Dr. Giacomo Carrea for having introduced me to the world of applied enzymology and for his helpful advice.

REFERENCES

1. Fieser, L. F., and M. Fieser (eds.) *Steroids*, Reinhold, New York, 1959.
2. Djerassi, C. (ed.) *Steroid Reactions, An Outline for Organic Chemists* Holden-Day, San Francisco, 1963.
3. Akhrem, A. A., and Titov, Y. A., (eds.). *Total Synthesis of Steroids,* IPST Press, Jersulalem, 1969.

4. Fried, J., and Edwards, J. A. (eds.) *Organic Reactions in Steroid Chemistry,* Reinhold, New York, 1972.

5. Blickenstaff, R. T., Ghosh, A. C. and Wolf, G. C., (eds.), *Total Synthesis of Steroids,* Academic Press, New York, 1974.

6. Charney, W., and Herzog, H. L. (eds.). *Microbial Transformation of Steroids,* Academic Press, New York, 1967.

7. Iisuka, H., and Naito, A. (eds.) *Microbial Conversion of Steroids and Alkaloids,* Springer-Verlag, New York, 1981.

8. Sandberg, E. C., and Jenkins, R. C., Hydrolysis of the sulfuric acid esters of steroid compounds, in *Methods in Enzymology,* Vol. 15 (R. B. Clayton, ed.), Academic Press, New York, 1969, pp. 684-690.

9. Roy, A. B., Evaluation of sulfate esters and glucuronides by enzymatic means, *Anal. Biochem. 165,* 1-12 (1987).

10. Chenault, H. K., and Whitesides, G. M., Regeneration of nicotinamide cofactors for use in organic synthesis, *Appl. Biochem. Biotechnol. 14,* 147-197 (1987).

11. Mosbach, K., (ed.), *Methods in Enzymology,* Vol. 136-138, Academic Press, New York, 1987 and 1988.

12. Riva, S., Bovara, R., Pasta, P., Carrea, G., Preparative scale regio- and stereospecific oxidoreduction of cholic acid and dehydrocholic acid catalyzed by hydroxysteroid dehydrogenases, *J. Org. Chem.,51,* 2902-2906 (1986).

13. Riva, S., Bovara, R., Zetta, L., Pasta, P., Ottolina, G., and Carrea, G., Enzymatic α/β inversion of C-3 hydroxyl of bile acids and study of the effects of organic solvents on reaction rates, *J. Org. Chem. 53,* 88-92 (1988).

14. Riva, S., Ottolina, G., Carrea, G., and Danieli, B., Efficient preparative scale enzymatic synthesis of specifically deutriated bile acids, *J. Chem. Soc. Perkin Trans. I,* 2073-2074 (1989).

15. Carrea, G., Bovara, R., Cremonesi, P., and Lodi, R., Enzymatic preparation of 12-ketochenodeoxycholic acid with NADP regeneration, *Biotechnol. Bioeng. 26,* 560-563 (1984).

16. Carrea, G., Bovara, R., Longhi, R., and Barani, R., Enzymatic reduction of dehydrocholic acid to 12-ketochenodeoxycholic acid with NADH regeneration, *Enzyme Microb. Technol., 6,* 307-311 (1984).

17. Carrea, G., Bovara, R., Longhi, R., and Riva, S., Preparation of 12-ketochenodeoxycholic acid from cholic acid using coimmobilized 12α-hydroxysteroid dehydrogenase and glutamate dehydrogenase with NADP cycling at high efficiency, *Enzyme Microb. Technol., 7,* 597-600 (1985).

18. Hofmann, A. F., The preparation of chenodeoxycholic acid and its glycine and taurine conjugates, *Acta Chem. Scand. 17,* 173-186 (1963).

19. Braun, M., Buckmann, A. F., and Riva, S., 12α-Hydroxysteroid dehydrogenase from *Clostridium* group P, strain C48-50: Production, purification, characterization and application," poster communication presented at the 10th Enzyme Engineering Conference, Kashikojima, Japan, 1989. To be published, *Annals of the New York Academy of Science.*

20. Buckman, A F., Carrea, G., Kulbe, K. D., Continuous synthesis of fine chemicals by cofactor dependent enzymes with simultaneous cofactor regeneration,

Biotechnology Action Programme (*BAP*), contract numbers 0060, 0065, and 0059.

21. Collins, D. C., Williamson, G., and Layne, D. S., Enzymatic synthesis by a partially purified transferase from rabbit liver microsomes, *J. Biol. Chem. 245* 873–876 (1970).

22. Adams, J. B., and McDonald, D., Enzymatic synthesis of steroid sulfates. XVI, Specificity and regulation of human adrenal hydroxysteroid sulfotransferase, *Steroids, 41* 575–586 (1983), and references therein.

23. Noguchi, S., Morita, K., and Nishikawa, M., Enzymatic hydrolysis of steroidal esters and its application to the syntheses of steroids, *Chem. Pharm. Bull. 8,* 563–565 (1960).

24. Wynne, K. N., and Renwick, A. G. C., The preparation of 16β-hydroxy-dehydroepiandrosterone using enzymatic deacylation, *Steroids, 19,* 293–300 (1972).

25. Rahim, M. A., and Sih, C. J., Microbial steroid esterases, in *Methods in Enzymology*, Vol. 15 (R. B. Clayton, ed.), Academic Press, New York, 1969, pp. 675–684

26. Neidleman, S. L., Diassi, P. A., Junta, B., Palmere, R. M., and Pan, S. C., The enzymatic halogenation of steroid, *Tetrahedron Lett. 44,* 5337–5342 (1966).

27. Levine, S. D., Neidleman, S. L., and Oberc, M., An enzymatic route to α-bromo sterodial ketones, *Tetrahedron,24,* 2979–2984 (1968).

28. Neidleman, S. L., and Levine, S. D., Enzymatic bromohydrin formation, *Tetrahedron Lett., 37,* 4057–4059, (1968).

29 For a recent review, see Walsh, C. T., and Chen, Y. C. J., Enzymatic Baeyer–Villiger oxidation by flavin-dependent monooxygenases, *Angew. Chem., Int. Ed. Engl., 27,* 333–343 (1988).

30. Prairie, R. L., and Talalay, P., Enzymatic formation of testololactone, *Biochemistry, 2,* 203–208 (1963).

31. Rahim, M. A., and Sih, C. J., Mechanisms of steroid oxidation by microorganisms, *J. Biol. Chem.241,* 3615–3623 (1966).

32. Veibel, S. Enzymatic synthesis of alkyl glucosides, *Enzymologia 1,* 124 (1936).

33. Ooi, Y., Hashimoto, T., Mitsuo, N., and Satoh, T., Enzymatic synthesis of chemically unstable cardiac glycosides by β-galctosidase from *Aspergillus oryzae*, *Tetrahedron Lett., 25,* 2241–2244 (1984).

34. Ooi, Y., Hashimoto, T., Mitsuo, N., and Satoh, T., Enzymic formation of β-alkyl glycosides by β-galctosidase from *Aspregillus oryzae* and its application to the synthesis of chemically unstable cardiac glycosides, *Chem. Pharm. Bull. 33,* 1808–1814 (1985).

35. Carrea, G., Biocatalysis in water–organic solvent two-phase systems, *Trends Biotechnol 2,* 102–106 (1984).

36. Lilly, M. D. , Harbron, S., and Narendranathan, T. J., Two-liquid-phase biocatacytic reactors, in *Methods in Enzymology,* Vol. 136, K. Mosbach, (ed.), Academic Press, New York, 1987, pp. 138–149.

37. Antonini, E., Carrea, G., and Cremonesi, P., Enzyme catalyzed reactions in water-organic solvent two-phase systems, *Enzyme Microb. Technol., 3,* 291–296 (1981).

38. For a recent review, see Carrea, G., and Cremonesi, P., Enzyme catalyzed transformations in water-organic solvent two-phase systems, in *Methods in Enzymology,* Vol. 136 (K. Mosbach, ed.), Academic Press, New York, 1987, pp. 150–157.

39. Cremonesi, P., Carrea, G., Ferreara, L., and Antonini, E., Enzymatic preparation of 20β-hydroxysteroids in a two-phase system, *Biotechnol. Bioeng. 17,* 1101–1108 (1975).

40. Cremonesi. P., Carrea, G., Sportoletti, G., and Antonini. E., Enzymatic dehydrogenation of steroids by β-hydroxysteroid dehydrogenase in a two-phase system, *Arch. Biochem. Biophys. 159,* 7–10 (1973).

41. Carrea, G., Riva, S., Bovara, R., and Pasta, P., Enzymatic oxidoreduction of steroids in two-phase systems: Effects of organic solvents on enzyme kinetics and evaluation of the performance of different reactors, *Enzyme Microb. Technol., 10,* 333–340 (1988).

42. Cremonesi, P., Mazzola, G., and Cremonesi, L., Enzyme-catalyzed reactions in a water–organic solvent heterogeneous system, *Ann. Chim. 67,* 415–422 (1977).

43. Lugaro, G., Carrea, G., Cremonesi, P., Casellato, M. M., and Antonini, E., The oxidation of steroid hormones by fungal laccase in emulsion of water and organic solvents. *Arch. Biochem. Biophys. 159,* 1–6 (1973).

44. For a recent review, see Luisi, P. L., and Laane, C., Solubilization of enzymes in apolar solvents via reversed micelles, *Trends Biotechnol, 4,* 153–161 (1986).

45. Hilhorst, R., Laane, C., and Veeger. C., Enzymatic conversion of apolar compounds in organic media using an NADH-regenerating system and dihydrogen as reductant, *FEBS Lett. 159,* 153–161 (1983).

46. Hilhorst, R., Sprujit, R., Laane, C., and Veeger, C., Rules for the regulation of enzyme activity in reversed micelles as illustrated by the conversion of apolar steroids by the 20β-hydroxysteroid dehydrogenase, *Eur. J. Biochem. 144,* 459–466 (1984).

47. Klibanov, A. M., Enzymatic catalysts in anhydrous organic solvents, *Trends Biochem. Sci. 14,* 141–144 (1989).

48. Klibanov, A. M., Enzymes that work in organic solvents, *Chemtech 16,* 354–359 (1986).

49. Zaks, A., and Klibanov, A. M., Enzyme catalyzed processes in organic solvents, *Proc. Natl. Acad. Sci. USA, 82,* 3192–3196 (1985).

50. Riva, S., and Klibanov, A. M., Enzymochemical regioselective oxidation of steroids without oxidoreductases, *J. Am. Chem. Soc. 110,* 3291–3295 (1988).

51. Riva, S., Bovara, R., Ottolina, G., Secundo, F., and Carrea, G., Regioselective acylation of bile acid derivatives with *Candida cylindracea* lipase in anhydrous benzene, *J. Org. Chem 54,* 3161–3164 (1989).

52. Njar, V. C. O., and Caspi, E., Enzymatic tranesterification of steroid esters in organic solvents, *Tetrahedron Lett. 28,* 6549–6552 (1987).

53. Randolph, T. W., Clark, D. S., Blanch, H. W., and Prausnitz, J. M., Enzymatic oxidation of cholesterol aggregates in supercritical carbon dioxide, *Science, 238, 387–390,* (1988).

54. Randolph, T. W., Clark, D. S., Blanch, H. W., and Prausnitz, J. M., Cholesterol aggregation and interaction with cholesterol oxidase in supercritical carbon dioxide, *Proc. Natl. Acad. Sci. USA, 85,* 2979–2983 (1988).

55. Lee, K. M., and Biellmann, J. F., Cholesterol conversion to Δ^4-cholestenone by cholestrol oxidase in polyphasic systems: Extension to the selective oxidation of 7β-hydroxycholesterol, *Tetrahedron, 44,* 1135–1139 (1988), and references therein.

56. Bergmeyer, H. U. (ed. in chief) *Methods of Enzymatic Analysis,* Vol. 8, 3rd ed., VCH, Weinheim, Germany, 1985, pp.139–316.

57. Iwata, T., and Yamasaki, K., Enzymatic determination and thin-layer chromatography of bile acids in blood, *J. Biochem, 56,* 424–431 (1964).

58. Mashige, F., Kazuhiro, I., and Osuga, T., A simple and sensitive assay of total serum bile acids, *Clin. Chim. Acta, 70,* 79–86 (1976).

59. Mashige, F., Tanaka, N., Maki, A., Kamei, S., and Yamanaka, M., Direct spectrophotometry of total bile acids in serum, *Clin. Chem. 27,* 1352–1356 (1981).

60. Styrelius, I., Thore, A., and Bjorkhem, I., Bioluminescent assay for total bile acids in serum with use of bacterial luciferase *Clin. Chem. 29,* 1123–1127 (1983).

61. Roda, A., Girotti, S., Ghini, S., and Carrea, G., Continous-flow assays with nylon tube-immobilized bioluminescent enzymes, in *Methods in Enzymology,* Vol. 137, (K. Mosbach, ed.), Academic Press, New York, 1988, pp. 161–171.

62. Kazandjian, R. Z., Dordick, J. S., and Klibanov, A. M., Enzymatic analyses in organic solvents, *Biotechnol. Bioeng. 28,* 417–421 (1986).

Index

Printed and bound by CPI Group (UK) Ltd, Croydon, CR0 4YY

17/10/2024

01775700-0009